Preserving

BY
THE EDITORS OF TIME-LIFE BOOKS

TIME-LIFE BOOKS/ALEXANDRIA, VIRGINIA

Cover: Boiling syrup is ladled into a jar of whole, uncooked cherries before the fruit is heat-processed *(Chapter 2)*. Plain sugar syrup would serve, but for extra color and flavor, the syrup here is made by adding small and misshapen cherries — and those too ripe for preserving whole — to the sugar and water before it is boiled. The syrup is then reduced and strained.

Time-Life Books Inc.
is a wholly owned subsidiary of
TIME INCORPORATED
Founder: Henry R. Luce 1898-1967
Editor-in-Chief: Henry Anatole Grunwald
President: J. Richard Munro
Chairman of the Board: Ralph P. Davidson
Executive Vice President: Clifford J. Grum
Chairman, Executive Committee: James R. Shepley
Editorial Director: Ralph Graves
Group Vice President, Books: Joan D. Manley
Vice Chairman: Arthur Temple

TIME-LIFE BOOKS INC.

Managing Editor: Jerry Korn; *Executive Editor:* David Maness; *Assistant Managing Editors:* Dale M. Brown (planning), George Constable, Thomas H. Flaherty Jr. (acting), Martin Mann, John Paul Porter; *Art Director:* Tom Suzuki; *Chief of Research:* David L. Harrison; *Director of Photography:* Robert G. Mason; *Assistant Art Director:* Arnold C. Holeywell; *Assistant Chief of Research:* Carolyn L. Sackett; *Assistant Director of Photography:* Dolores A. Littles; *Production Editor:* Douglas B. Graham; *Operations Manager:* Gennaro C. Esposito, Gordon E. Buck (assistant); *Assistant Production Editor:* Feliciano Madrid; *Quality Control:* Robert L. Young (director), James J. Cox (assistant), Daniel J. McSweeney, Michael G. Wight (associates); *Art Coordinator:* Anne B. Landry; *Copy Staff:* Susan B. Galloway (chief), Nancy Berman, Tonna Gibert, Celia Beattie; *Picture Department:* Alvin Ferrell; *Traffic:* Kimberly K. Lewis

Chairman: John D. McSweeney; *President:* Carl G. Jaeger; *Executive Vice Presidents:* John Steven Maxwell, David J. Walsh; *Vice Presidents:* George Artandi (comptroller); Stephen L. Bair (legal counsel); Peter G. Barnes; Nicholas Benton (public relations); John L. Canova; Beatrice T. Dobie (personnel); Carol Flaumenhaft (consumer affairs); James L. Mercer (Europe/South Pacific); Herbert Sorkin (production); Paul R. Stewart (marketing)

THE GOOD COOK

The original version of this book was created in London for Time-Life International (Nederland) B.V.
European Editor: Kit van Tulleken; *Design Director:* Louis Klein; *Photography Director:* Pamela Marke; *Chief of Research:* Vanessa Kramer; *Special Projects Editor:* Windsor Chorlton; *Chief Sub-Editor:* Ilse Gray; *Production Editor:* Ellen Brush; *Quality Control:* Douglas Whitworth

Staff for *Preserving:* *Series Editor:* Alan Lothian; *Series Coordinator:* Liz Timothy; *Head Designer:* Rick Bowring; *Text Editor:* Ann Tweedy; *Anthology Editor:* Markie Benet; *Staff Writers:* Alexandra Carlier, Jay Ferguson, Ellen Galford, Thom Henvey; *Designer:* Zaki Elia; *Researchers:* Nora Carey, Sally Crawford, Deborah Litton; *Sub-Editors:* Kathy Eason, Katie Lloyd, Sally Rowland; *Design Assistants:* Sally Curnock, Ian Midson; *Editorial Department:* Kate Cann, Debra Dick, Beverley Doe, Philip Garner, Aquila Kegan, Lesley Kinahan, Debra Lelliott, Linda Mallett, Molly Sutherland, Julia West, Helen Whitehorn

U.S. Staff for *Preserving:* *Series Editor:* Gerry Schremp; *Assistant Editor:* Ellen Phillips; *Designer:* Ellen Robling; *Chief Researcher:* Barbara Fleming; *Picture Editor:* Christine Schuyler; *Text Editor:* Mark Steele; *Staff Writers:* Fran Moshos, David Schwartz; *Researchers:* Cynthia Jubera (techniques), Karin Kinney (anthology); *Assistant Designer:* Peg Schreiber; *Art Assistant:* Robert Herndon; *Editorial Assistants:* Brenda Harwell, Patricia Whiteford

CHIEF SERIES CONSULTANT

Richard Olney, an American, has lived and worked for some three decades in France, where he is highly regarded as an authority on food and wine. Author of *The French Menu Cookbook* and of the award-winning *Simple French Food*, he has contributed also to numerous gastronomic magazines in France and the United States, including the influential journals *Cuisine et Vins de France* and *La Revue du Vin de France*. He is a member of several distinguished gastronomic societies, including La Confrérie des Chevaliers du Tastevin, Les Amitiés Gastronomiques Internationales, and La Commanderie du Bontemps de Médoc et des Graves. Working in London with the series editorial staff, he has been basically responsible for the planning of this volume, and has supervised the final selection of recipes submitted by other consultants. The United States edition of The Good Cook has been revised by the Editors of Time-Life Books to bring it into complete accord with American customs and usage.

CHIEF AMERICAN CONSULTANT

Carol Cutler is the author of a number of cookbooks, including the award-winning *The Six-Minute Soufflé and Other Culinary Delights*. During the 12 years she lived in France, she studied at the Cordon Bleu and the École des Trois Gourmandes, and with private chefs. She is a member of the Cercle des Gourmettes, a long-established French food society limited to just 50 members, and a charter member of Les Dames d'Escoffier, Washington Chapter.

SPECIAL CONSULTANTS

Dora Jonassen learned about preserving as a child on a farm in her native Denmark. She received a degree in Home Economics from Sorö College and taught cooking and nutrition before moving to New York in 1963. A food stylist, writer and consultant, she has been responsible for most of the step-by-step photographic sequences in this volume.
Ruth Klippstein, author of "Drying Foods at Home" and "Freezing Handbook," has a degree in Nutrition from Michigan State. Professor of Nutritional Sciences at Cornell University, she heads the food preservation program of the U.S. Department of Agriculture, Cooperative Extension for New York.

PHOTOGRAPHER

Aldo Tutino has worked in Milan, New York City and Washington, D.C. He has received a number of awards for his photographs from the New York Advertising Club.

INTERNATIONAL CONSULTANTS

GREAT BRITAIN: *Jane Grigson* has written a number of books about food and has been a cookery correspondent for the London *Observer* since 1968. *Alan Davidson*, a former member of the British Diplomatic Service, is the author of several cookbooks. *Pat Alburey* is a member of the Association of Home Economists of Great Britain. She has been responsible for some of the step-by-step photographic sequences in this volume. *Jean Reynolds*, who preserved some of the food for the photographs in this volume, is from San Francisco. She trained as a cook in the kitchens of several of France's great restaurants. FRANCE: *Michel Lemonnier*, the co-founder and vice president of Les Amitiés Gastronomiques Internationales, is a frequent lecturer on wine and vineyards. GERMANY: *Jochen Kuchenbecker* trained as a chef, but worked for 10 years as a food photographer in several European countries before opening his own restaurant in Hamburg. *Anne Brakemeier* is the co-author of a number of cookbooks. ITALY: *Massimo Alberini* is a well-known food writer and journalist, with a particular interest in culinary history. His many books include *Storia del Pranzo all'Italiana*, *4000 Anni a Tavola* and *100 Ricette Storiche*. THE NETHERLANDS: *Hugh Jans* has published cookbooks and his recipes appear in several Dutch magazines. THE UNITED STATES: *Judith Olney*, author of *Comforting Food* and *Summer Food*, received her culinary training in England and France. In addition to conducting cooking classes, she regularly writes articles for gastronomic magazines. *Robert Shoffner*, wine and food editor of *The Washingtonian* magazine since 1975, has written many articles on cuisine.

Correspondents: Elisabeth Kraemer (Bonn); Margot Hapgood, Dorothy Bacon, Lesley Coleman (London); Susan Jonas, Lucy T. Voulgaris (New York); Maria Vincenza Aloisi, Josephine du Brusle (Paris); Ann Natanson (Rome).
Valuable assistance was also provided by: Janny Hovinga (Amsterdam); Judy Aspinall, Karin B. Pearce (London); Bona Schmid (Milan); Carolyn T. Chubet, Miriam Hsia, Christina Lieberman (New York); Mimi Murphy (Rome).

For information about any Time-Life book, please write:
Reader Information, Time-Life Books
541 North Fairbanks Court, Chicago, Illinois 60611

Library of Congress CIP data, page 176.

CONTENTS

The Art of Fixing the Seasons

Nothing in the realm of cookery comes closer to alchemy than preserving. In the natural course of things, food quickly spoils. But if it is subjected to extremes of cold or heat, or to a high concentration of salt, sugar, vinegar or alcohol, its normal decay can be arrested. With appropriate precautions, the food will keep for weeks, months, even years.

The discovery of preserving techniques freed the human diet from bondage to the seasonal cycle, in which the plenty of spring and summer was followed by months of meager fare. The 19th Century French gastronome Grimod de la Reynière was moved to call preserving "the art of fixing the seasons" and to rhapsodize over the way it enables the cook to "recall the month of May in the heart of winter."

Today, thanks to a sophisticated produce-farming industry with an extensive transportation network, it is possible to buy almost any kind of produce anywhere at any time of year. But the brief abundance from local farms and home gardens is certainly less expensive. Preserving, therefore, has economy to recommend it. Another compelling reason for preserving is that foods produced locally and harvested at the height of perfection have a flavor superior to that of any specimen picked when underripe and shipped thousands of miles. Home-grown August tomatoes canned in their own juice *(pages 32-33)* and a lightly cooked jam of summer peaches *(recipe, page 106)* are more reminiscent of sun-warmed vegetables and fruits than the wan products of intensive and remote cultivation.

Some approaches to preserving are not intended to retain the original character of food: These processes transform flavor and texture in such a way as to to give results that many people rate higher than the fresh products. The ordinary methods of everyday cookery cannot achieve, for example, the tart savor of old-fashioned dill pickles *(pages 86-87),* the rich taste of smoked fish *(page 90),* or the complex blend of flavors in the pear, citrus fruit and nut conserve whose elements are shown opposite *(demonstration, pages 56-57).*

This volume offers a comprehensive guide to the manifold pleasures and possibilities of preserving food at home. It begins by examining the equipment required for the two most frequently used preserving techniques: freezing and canning. The survey of equipment is followed by five chapters of step-by-step demonstrations of the basic preserving methods: freezing; canning (sterilizing with heat); the making of jellies, jams and other fruit mixtures preserved with sugar; preserving in vinegar or alcohol; and preserving by extracting the moisture from foods. With the knowledge gained from these practical lessons, you will be able to create your own unique preserves, as well as prepare any of the 250 recipes in the anthology that makes up the second half of this volume.

Why food spoils

As inevitably as foods mature, so do they subsequently deteriorate. Their loss of quality has two principal causes: enzymes, which are proteins present in all animal and vegetable tissues, and microorganisms, which are the molds, yeasts and bacteria found in all fresh foods.

Enzymes are catalysts, promoting the myriad chemical reactions that occur in living things. Long after fruits and vegetables have been harvested, or animals have been killed, the enzymes in their cells remain active, eventually changing the food's flavor, color and texture.

The microscopic fungi known as molds, present everywhere in the air, usually appear on foods as fine threads, blue, white or pale green in color. Molds have their uses—forming the veins in blue cheeses, for example. But for the most part, molds produce chemical substances that have unpleasant flavors and may even be poisonous. Furthermore, molds can reduce the acidity of foods—and acidity acts as a safeguard against other organisms that cause spoilage.

Yeasts, also ubiquitous in the air, feed on the sugars and starches in foods, from which yeasts produce alcohol and carbon dioxide and then cause fermentation. In some foods, yeast fermentation is desirable: Yeasts turn grape juice into wine and barley mash into beer and make bread dough rise. But more often, yeasts merely produce an unappealing sourness.

Bacteria, present in soil and water as well as the air, are the most pernicious spoilers of food; a number of them manufacture substances harmful to humans. Toxins produced by bacteria of the genus *Salmonella* and the species *staphylococcus aureus* cause intestinal disorders, and a substance made by the species *clostridium botulinum* causes the life-threatening—but fortunately rare—food poisoning known as botulism, which acts on the nervous system.

All these spoilage organisms are effectively destroyed or rendered harmless by the preserving techniques demonstrated in this book; the instructions for the preparation and sterilization of food should be scrupulously followed for safety's sake.

Arresting decay with heat and cold

Our knowledge of enzymes and microorganisms is fairly recent. Enzymes were discovered in experiments that spanned most of the 19th Century; a pure enzyme was not isolated until 1926. What we know about microorganisms stems largely from the work of the 19th Century French chemist and biologist Louis Pasteur. In the 1860s, Pasteur began studying fermentation in milk, vinegar, wine and beer, and showed that the process was caused by microorganisms. He went on to demonstrate that any microorganism's activity—be it fermentation or putrefaction—could be retarded or completely halted by subjecting food to heat, the short-term sterilizing treatment that came to be known as pasteurization.

Freezing is a technique so closely associated with sophisticated machinery that it might seem to be a modern development, but its roots lie deep in the past. Hundreds of thousands of years ago, people who lived in cold climates must have observed that frozen meat did not decay, and they may well have buried food below ice and snow to store it for later use. The Romans took typically bold advantage of the preservative effects of low temperatures. During winter, they transported ice from the Alps to southern cities and put it in caves or under woolen blankets or straw to prevent melting; in summer, they used the ice to keep food fresh (and also to cool the wine that they quaffed in such quantities). The practice of storing winter's ice for summer use lingered well into the 20th Century in many parts of the world.

Reducing the temperature of food to near the freezing point merely slows the activity of destructive microorganisms within the food. A deep freezer, however, keeps food at a temperature below freezing and thereby prevents the growth or production of any toxins. However, enzyme activity, although slowed considerably, is not entirely stopped, and foods still gradually deteriorate in the freezer. (Recommended storage times for fruits, vegetables, fresh meats, variety meats, fish and shellfish, and some prepared foods are given in the guide on pages 14-15.)

Unlike the cold of freezing, the extreme heat applied in canning completely halts the activities of enzymes as well as microorganisms. And in canning, heat not only sterilizes food but also seals its container, protecting the food from airborne contaminants. The mechanism of sealing is simple. Air in canning jars expands during heating, and some of it is forced out of the jars. When the jars cool, the remaining air contracts, creating a partial vacuum that pulls the lids tightly shut and holds them there. This technique was discovered in the early 19th Century by a Frenchman, Nicholas-François Appert, who succeeded in preserving a wide range of foods by heating them (the technical term for this is processing) in jars covered with a combination of corks, wire and wax. His jars were quickly succeeded by the simpler canning jars developed in 1858 by an American, John Mason; the jars in use today (pages 10-11) are descendants of Mason's invention and commonly bear his name.

How much heat is required to preserve food after it is packed in jars depends on the nature of the food itself. Acid in food, for instance, protects the food from enzyme deterioration and from many microorganisms. High-acid foods—tomatoes and all

An Overview of Preserving Possibilities

How to use the chart. Locate the food you plan to preserve in the row at the top of the chart. Choose a method of preserving the food from the possibilities shown by dots in the column beneath each entry. The eight methods listed are explained fully in the chapters that follow.

Vegetables

Preserving Method	Artichoke Bottoms	Asparagus	Beans: Broad/Lima	Beans: Green/Wax	Beets	Broccoli	Brussels Sprouts	Cabbage	Carrots	Cauliflower	Celery	Corn	Cucumbers	Eggplant	Fennel	Greens: Beet/Collard/Turnip	Herbs	Mushrooms	Okra	Olives	Onions	Parsnips	Peas: Sweet	Peppers: Chili	Peppers: Sweet	Potato: Sweet	Potato: White	Pumpkin	Rutabaga	Snow Peas	Sorrel	Spinach	Squash: Summer	Squash: Winter	Swiss Chard Leaves	Tomatoes	Turnips
Freezing		●	●	●		●	●		●	●		●				●	●	●	●				●	●	●	●		●	●	●	●	●	●	●	●		●
Canning	●		●	●					●			●											●			●										●	
Making Jellies and Jams														●										●	●	●								●		●	
Pickling and Alcohol Preserving	●	●	●	●	●	●	●	●	●	●	●	●	●	●	●			●	●	●	●	●	●	●	●	●										●	●
Fat-sealing																																					
Drying			●	●								●					●	●			●			●	●		●									●	
Salting			●	●	●	●	●	●	●	●						●	●	●	●																	●	●
Smoking																																					

fruits—are effectively preserved if they are heated to 212° F. [100° C.] in a bath of boiling water *(pages 32-39)*.

Other foods that may be processed in a boiling water bath include those partially protected from decay by high concentrations of sugar, vinegar (acetic acid) or alcohol. To this group belong the delectable, sweet preparations beloved by generations of industrious cooks: marmalades, jams, conserves and butters *(pages 46-59)*. Additional members of the group are seasoned vinegar pickles, enjoyed since ancient times. The word "pickle" is usually associated with cucumbers, but any firm vegetable or fruit can be pickled, alone or in the combinations that produce relishes, piccalillis and chutneys *(pages 62-73)*. As for foods that receive part of their protection from alcohol, perhaps the best-known is mincemeat, a mixture of chopped fruit, brandy, suet, spices and often meat *(pages 72-73)*. Mincemeat originated during the Middle Ages as a method of preserving meat after fall slaughtering, and it has come to be a culinary symbol of Christmas.

Foods low in natural acid or preserved without sugar, vinegar or alcohol require a higher temperature than that of a boiling water bath; these foods are processed in pressurized equipment *(pages 40-41)* that can raise the temperature of water to 240° F. [116° C.], the level needed to inactivate their enzymes and kill any bacteria they contain.

Without the aid of freezing or canning, foods can be preserved by salting, drying, smoking and sealing in fat, all of which deprive microorganisms of the moisture they need in order to grow. Salt dissolves in the food's liquid and makes it unusable for microorganisms, drying and smoking simply remove moisture, and sealing in fat protects the food from outside moisture and air.

The pleasures of preserving
The preserving procedures followed today have been in use, in one form or another, for generations. They carry with them a wealth of comforting associations—which is part of the motivation behind this form of cooking. The sight of brightly hued rows of jam jars and pickle jars adorning a pantry shelf, or the heady aroma of fish being smoked over an applewood fire, still evokes a sense of plenty.

Almost every edible object is material for preserving. The violets of May can serve to scent a vinegar; the hips—seed pods—of just-blown wild roses make the most delicate of jellies. High-summer crops such as peaches, peppers, tomatoes or corn yield sugary jams or tart pickles or in some cases fragrant liqueurs. Even autumn and winter have their bounty: Apples and chestnuts lend themselves to delectable preserves. Preserving is thus the most provident of cooking arts and, at the same time, it provides culinary adventures that no purchased preparation can ever successfully emulate.

	Fruits																							Meats							Variety Meats						Fish and Shellfish			Preserving Method
Apples	Apricots	Blackberries	Blueberries	Cherries	Citrus Fruits	Cranberries	Currants	Elderberries	Figs	Gooseberries	Grapes	Guavas	Melons	Nectarines	Peaches	Pears	Pineapple	Plums	Quinces	Raspberries	Rhubarb	Strawberries	Beef	Game Animals	Game Birds	Lamb	Pork	Poultry	Veal	Meat: Ground	Bone Marrow	Giblets	Heart	Kidney	Liver	Tongue	Fish	Crustaceans	Mollusks	
•	•	•	•	•	•	•	•	•	•	•	•	•	•	•	•	•	•	•	•	•	•	•	•	•	•	•	•	•	•	•	•	•	•	•	•	•	•	•	•	Freezing
•	•		•	•	•	•	•			•	•			•	•	•		•				•	•	•	•	•	•	•												Canning
•	•	•	•	•	•	•	•	•	•	•	•	•		•	•	•	•	•	•	•	•	•																		Making Jellies and Jams
•	•	•	•	•			•	•			•	•		•	•	•	•	•	•		•	•															•			Pickling and Alcohol Preserving
																							•	•	•	•	•	•	•						•	•	•			Fat-sealing
•	•			•	•		•		•		•			•	•	•	•	•				•	•	•	•												•			Drying
																							•	•	•	•	•	•	•						•	•	•			Salting
																							•	•	•	•	•	•									•		•	Smoking

Containers and Wrappings for Frozen Food

To be stored in the freezer, food must be shielded from the air: Air dries food, leaching its color and flavor, and toughening its texture. In the flesh of frozen meat, poultry and fish, air causes white stringy patches, a phenomenon known as freezer burn. Air also interacts with fat in frozen food, fostering rancidity.

Airtight packaging with the proper materials prevents these mishaps. Liquids and small foods can be packed in canning jars *(pages 10-11)* or in tight-lidded plastic freezer containers, both of which are widely available. To minimize air space in the containers, pack solid foods firmly up to the rims: Liquids need a 1-inch [2½-cm.] space beneath the rim to allow for expansion during freezing.

In many cases, the size or shape of food will preclude freezing in containers; instead, the food must be stored in bags or wrapped in packages made of impermeable material. Wax paper, regular aluminum foil, butcher paper and ordinary plastic wrap are all permeable and therefore unsuitable.

There are, however, many varieties of bags and wrapping materials designed for the freezer and labeled accordingly. The most effective of these is plastic-backed freezer paper, which is sold in rolls. Heavy-duty aluminum foil, also sold in rolls, can be used, but it may develop tiny holes when folded, admitting air. Clear plastic freezer wrap, sold as bags or in rolls, allows you to see the frozen food, but it becomes rigid when frozen; and unless the edges of the package are carefully taped, they will separate and let in air.

Two basic wrapping techniques are used to package food to be frozen. Large sturdy foods such as roasts may be rolled up in a butcher wrap *(top demonstration)*. To minimize handling, more fragile foods should be packaged by a drugstore wrap *(bottom demonstration)*. In either case, always press as much air as you can from the food package. Seal it with freezer or masking tape, both of which adhere well at low temperatures. Label any package, indicating its contents, the quantity of food within, and the date of freezing. Be sure to use the food within the time limit specified in the guide on pages 14-15.

The Butcher Wrap

1 **Positioning the food.** Lay a piece of freezer paper three times as large as the food to be wrapped shiny side up on a work surface. Place the food — here, a pork roast — diagonally across one corner of the paper so that the paper extends beyond both ends of the food. Pull the corner up and around the meat, tucking it underneath.

2 **Folding in the edge.** Fold the edge of paper at one end of the food over the food, and hold the paper firmly in place while you fold the other edge over the opposite end of the food.

The Drugstore Wrap

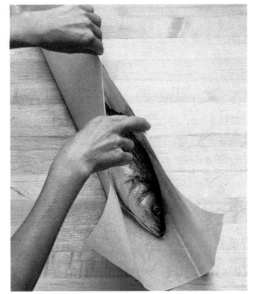

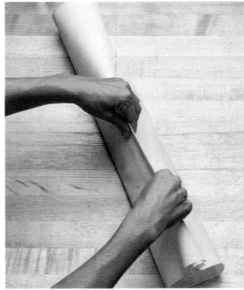

1 **Positioning the food.** Place a piece of freezer paper four times larger than the food to be wrapped shiny side up on a work surface. Lay the food — in this case, a glazed striped bass *(page 27)* — in the center of the paper. Lift the opposite edges of the paper up until they meet over the food.

2 **Folding the edges.** Fold the edges of the paper together, making a 1-inch [2½-cm.] fold. Turn this fold repeatedly over on itself until it rests snugly on top of the food. Gently press the fold to make a sharp crease that will prevent unrolling.

3 **Rolling the food.** Smooth the paper-covered food with your hands to expel air. Then, holding the paper in place, roll the food toward the opposite corner of the paper, pressing firmly to expel air from the package.

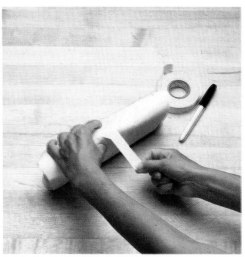

4 **Taping the package.** When the food is completely rolled up in paper, hold the free corner of paper against the package and secure it in place with a length of freezer tape.

5 **Labeling the package.** With an indelible pen, write on the package the type of food, the amount and the date of freezing. Freeze immediately.

3 **Closing one end.** Fold one end of the package to form a V-shaped tab. Fold this tab up over the package. Secure it in place with freezer tape.

4 **Closing the other end.** Run your hands over the package to expel air. Then fold the other end of the paper into a V-shaped tab, fold it over the package and tape it in place.

5 **Sealing securely.** Seal the creased fold on top of the package with freezer tape. Using an indelible pen, label the package with the name of the food, the amount and the date of freezing. Freeze immediately.

A Choice of Canning Jars

Canning means packing food into containers and heating—or processing—it to neutralize spoilage agents and create a partial vacuum that seals the container lids. Safe canning requires proper containers, carefully prepared. Do not use leftover jars such as peanut-butter jars: These probably cannot take the heat of processing, whether done in a boiling water bath or in the hotter environment of a pressure canner. Instead, use jars specially produced for home canning.

For either boiling-water-bath or pressure processing, the most convenient jar is made of heat-tempered glass and is fitted with a flat lid and a separate screw band that holds the lid in place. These jars range in size from ½ pint [¼ liter] to 1 quart [1 liter]; some have wide mouths for large pieces of food. (Half-gallon jars should not be used for canning; heat penetrates the contents too slowly.) Both jars and screw bands are reusable; the flat lids—sold separately—are not.

The second recommended type of container is a glass jar with a hinged glass lid that is clamped in place. For a tight seal, a rubber gasket fits between the jar and lid. This less sturdy type should be used only in a boiling water bath. The gaskets must be replaced for each use.

Either type of jar must be free of cracks or nicks to ensure sealing. After cleaning both jars and lids, leave them in clean hot water until you are ready to use them. Once filled, they should be processed at once, following the steps outlined for each type on pages 30-31.

The only foods canned without processing are jellies (pages 46-47). They may be poured hot into canning jars; the filled jars should be covered, then inverted for five seconds so the hot jelly can destroy organisms on the lids. Or jellies can be poured into everyday jars and covered with paraffin wax—available at supermarkets—instead of lids. Because they are not processed, jelly jars must be sterilized before being filled: Cover them with water and boil for 10 minutes.

Simple Screw-on Tops

1 **Filling jars.** Wash jars, lids and screw bands. Leave the jars in hot water; put the lids and bands in a pan of water, bring to a boil, then remove from the heat. Before use, drain the jars on a towel. Fill each jar—here, with tomatoes. Run a nonmetallic utensil around the inside of the jar to expel air.

Using Clamped Glass Lids

1 **Fitting on gaskets.** Wash and drain the jars (Step 1, above, right). Wash the gaskets and place them in a pan; cover with simmering, not boiling, water. Then fit a gasket onto the rim of each jar lid.

2 **Securing the lids.** Fill the jars—here, with peach jam—and expel excess air (Step 1, above, right). Clamp the lids shut. Process the jars; when processing time is up, let the jars cool to room temperature in the water. Then remove the jars from the canning vessel and let them cool an additional 12 hours.

3 **Testing seal.** Gently unlock the clamp of each cooled jar. Holding the jar by the lid, lift it very slightly. If the jar remains closed, a seal has formed and the jar may be reclamped and stored. If not, reprocess the jar using a new gasket, or refrigerate the jar and eat the contents within a few days.

2 **Covering jars.** With a piece of muslin or cheesecloth dipped in hot water, wipe the top and sides of the jar rim clean: Food particles will interfere with sealing. Remove a lid from the hot water and fit it onto the top of the jar.

3 **Processing.** Fit a screw band snugly over the jar rim and lid. Do not force a tight fit: Air must escape in processing. Process the jars. When the time is up, use tongs to transfer the jars to a draft-free place. Cool for at least 12 hours. Popping sounds from the jars indicate that seals are forming.

4 **Testing the seal.** When cooled jars have sealed, the lids are concave. Press each lid; it should remain concave. If so, remove the screw bands and store the jars in a cool, dark place. If not, either reprocess the jars with new lids or refrigerate them and eat the contents within a few days.

A Covering of Paraffin

1 **Melting paraffin.** Place chunks of paraffin wax in an uncovered, heatproof —and disposable—container; a clean soup can is used here. Place the container in a pan of water and leave over low heat, stirring gently until the paraffin has melted. Never melt the paraffin over direct heat; it might ignite.

2 **Covering jelly.** Sterilize jars, drain and fill to ½ inch [1 cm.] of the rims with hot jelly—in this case, grape jelly. Wipe the inside edges of the jars clean. Immediately pour hot paraffin over the jelly to a depth of ⅛ inch [3 mm.]. With a needle, prick any bubbles that appear on the paraffin. Set the jars in a draft-free place to cool.

3 **Forming the seal.** As the paraffin cools, it will harden, become opaque, and sink to form a concavity. If not— or if jelly seeps out from under the edges of the paraffin—remove the paraffin and repeat the entire sealing process. To remove paraffin, slide the tip of a small knife down the inside edge of the jar and pry up the paraffin cover.

1

Freezing
Suspending Natural Processes

Blanching in liquid or steam
Molding gratins and purées
Maintaining texture with sugar
Preliminaries for poultry and meat
A sheath of ice for fish

In preparation for freezing, a fresh chicken, cleaned and trussed (pages 22-23), is enclosed in plastic freezer wrap. The parcel will be wrapped in heavy-duty foil and labeled with the freezing date so that the chicken can be used while it is still in prime condition.

Of all home preserving methods, freezing—reducing the temperature of food to 0° F. [-18° C.]—is the easiest and the one that keeps the food closest to its original form. Success requires only that you follow a few simple rules.

Food to be frozen should be in prime condition: Vegetables and fruits, for instance, should be frozen soon after they are picked, before their quality fades. During preparation the food can be contaminated by bacteria that will harm quality or cause decay when the food is thawed; to minimize this, keep utensils and work surfaces scrupulously clean.

Freezing drastically reduces, but does not stop, the activity of enzymes—proteins that stimulate aging changes in all food. Enzyme activity does not harm frozen meats or fish and is neutralized by the acids in frozen fruits. However, all of the vegetables that are suitable for freezing are low in acid and most require a brief preliminary cooking *(pages 16-17)* to halt enzyme action.

All food to be frozen must be carefully shielded from the air by the wrappings and containers described on pages 8-9. After these preliminaries, the only requirement for successful preserving is an efficient freezer—one whose temperature stays steadily at or below 0° F. During freezing, the water naturally present in food expands as it turns into ice crystals. Because crystals can disrupt the food's cells, a freezer should freeze foods quickly, thereby minimizing the size of the crystals and the damage they do. In addition to following the manufacturer's instructions for operating the freezer, monitor the temperature with a freezer thermometer, available at kitchen-equipment shops.

No matter how carefully you prepare food and how efficient your freezer, the food's structure will be broken down to some extent, exposing ruptured cells to bacterial action when the food is thawed. The longer the storage time, the greater the damage. To protect the food, use it within the time periods specified in the guide on pages 14-15, and thaw it either in very hot or in cold conditions, where microorganisms will not thrive. Vegetables are quickly penetrated by heat, and excellent results are obtained by cooking them directly from a frozen state. Fruits to be eaten raw, as well as meats and fish, should be thawed slowly in the refrigerator—and used as soon as they have thawed.

The Fundamentals of Freezing

Temperature is the key to effective freezing. Any home freezer or freezing compartment must be capable of bringing the temperature of fresh food to 0° F. [-18° C.] or lower and holding it there. Ice-making compartments are not cold enough: They should never be used to freeze foods.

Your freezer's instruction booklet will tell you how to operate the controls and how much food can be frozen at one time. Most models will freeze up to 10 per cent of their capacity in 24 hours—that is, about 2½ pounds per cubic foot [1¼ kg. per 28 liters]. Usually there is no need to adjust the temperature for food weighing 2½ pounds or less. However, to freeze more than that, reduce the temperature several hours in advance. If there is a fast-freeze switch, use it; otherwise, turn the thermostat to its lowest setting.

The freezer instruction booklet also will explain how to pack food into the freezer. Once the food is frozen, a freezer thermometer will enable you to check that your freezer is maintaining the correct temperature.

To keep your freezer functioning efficiently, defrost it regularly (unless, of course, you have a frost-free model).

A properly operating freezer can preserve almost any food. Exceptions are few: Whole eggs expand, cracking their shells; the whites of cooked eggs turn rubbery. Emulsions such as mayonnaise and any dishes or sauces thickened with egg or starch will curdle or separate.

Almost all other fresh foods and cooked dishes can be frozen, although some will fare better than others: Foods that freeze least well are those containing fat, which can quickly become rancid.

The guide below lists foods most suitable for freezing and indicates the methods—demonstrated on the pages that follow—for preparing and packing them.

Before packing, all foods should be prepared as for eating or cooking: Vegetables and fruits should be peeled and seeded as necessary, and cut into pieces; meats and fish should be trimmed into the forms in which they will be cooked.

Other preparations noted in the guide are designed to maintain the food in the best condition possible, protecting it from the drying freezer air and from changes wrought by its own chemistry.

In the case of fruits, for instance, added sugar safeguards texture and flavor. The sugar may be added in one of three ways, depending on the fruit (pages 20-21). Some fruits can be dry-packed—layered with sugar in airtight containers. Some fruits are suitable for tray-freezing: They

A Guide to Times and Treatments

Vegetables

Asparagus. Steam-blanch tips for two minutes, stalks for three minutes. Tray-freeze or dry-pack. Storage life: eight to 12 months.

Beans, broad and lima. Water-blanch for two minutes. Tray-freeze or dry-pack. Storage life: 10 to 12 months.

Beans, green and wax. Water-blanch for two minutes. Tray-freeze or dry-pack. Storage life: eight to 12 months.

Broccoli. Soak buds for 30 minutes in salted water to remove insects; steam-blanch for five minutes. Tray-freeze or dry-pack. Steam-blanch stalks for five minutes, and tray-freeze or dry-pack. Storage life: 10 to 12 months.

Brussels sprouts. Soak for 30 minutes in salted water to remove insects; water-blanch for four minutes. Tray-freeze or dry-pack. Storage life: 10 to 12 months.

Carrots. Water-blanch pieces ½ inch [1 cm.] thick for three minutes. Tray-freeze or dry-pack. Storage life: 10 to 12 months.

Cauliflower. Soak florets for 30 minutes in salted water to remove insects; steam-blanch for five minutes. Tray-freeze or dry-pack. Storage life: 10 to 12 months.

Corn kernels. Water-blanch for four and one half minutes on cob; cut kernels from cob. Tray-freeze or dry-pack. Storage life: 10 to 12 months.

Corn on the cob. Water-blanch for eight minutes. Dry-pack individually. Storage life: 10 to 12 months.

Greens (beet, mustard, turnip). Steam-blanch for one and one half minutes. Dry-pack. Storage life: 10 to 12 months.

Herbs. Dry-pack. Storage life: six months.

Kale. See Greens.

Mushrooms. Sauté for three minutes in butter or oil. Wet-pack. Storage life: 10 to 12 months.

Okra. Water-blanch for three minutes. Tray-freeze or dry-pack. Storage life: 10 to 12 months.

Onions. Dry-pack. Storage life: 10 to 12 months.

Parsnips. Shred or grate; sauté in butter. Wet-pack. Storage life: eight to 10 months.

Peas, sweet. Water-blanch for one minute. Tray-freeze or dry-pack. Storage life: 10 to 12 months.

Peppers. Dry-pack; or broil until skins are charred, peel and wet-pack. Storage life: 10 to 12 months.

Pumpkin. Bake whole in a 350° F. [180° C.] oven until tender when pierced with a fork (at least one hour). Remove pulp and dry-pack. Storage life: 10 to 12 months.

Rutabaga. Shred or grate; sauté in butter. Wet-pack. Storage life: eight to 10 months.

Snow peas. Water-blanch for one minute. Tray-freeze or dry-pack. Storage life: 10 to 12 months.

Sorrel. See Greens.

Spinach. See Greens.

Squash, summer. Water-blanch for one minute. Tray-freeze or dry-pack. Storage life: 10 to 12 months.

Squash, winter. See Pumpkin.

Swiss chard leaves. See Greens.

Turnips. Shred or grate; sauté in butter. Wet-pack. Storage life: eight to 10 months.

Fruits

Apples. Steam-blanch for one and one half minutes. Dry-pack with sugar. Storage life: 10 to 12 months.

Apricots. Wet-pack with acidulated syrup. Storage life: 10 to 12 months.

Berries (except strawberries). Tray-freeze or dry-pack. Storage life: 10 to 12 months.

Cherries. Dry-pack with sugar for use in pies, or tray-freeze. Storage life: 10 to 12 months.

Currants. See Berries.

Figs. Dry-pack with sugar. Storage life: 10 to 12 months.

Grapefruit. Wet-pack. Storage life: 10 to 12 months.

Grapes. Tray-freeze or dry-pack. Storage life: 10 to 12 months.

Melons. Wet-pack. Storage life: 10 to 12 months.

Nectarines. Wet-pack with acidulated syrup. Storage life: 10 to 12 months.

Oranges. See Grapefruit.

are frozen on a baking sheet in a single, sugar-sprinkled layer so that they do not stick together, then are packed loosely in airtight containers, from which any amount of fruit can be removed.

Some fruits must be wet-packed—covered with a liquid such as sugar syrup to seal them from the air. This treatment helps protect the colors of pale-fleshed fruits such as peaches. For additional insurance against discoloration, the sugar syrup for these fruits is augmented with lemon juice or ascorbic acid—widely available vitamin C crystals. To each quart of syrup, add 1 teaspoon [5 ml.] of the juice or ½ teaspoon [2 ml.] of the crystals, dissolved in a little water.

Vegetables, unlike fruits, are generally low in acid, and require brief cooking before freezing to inactivate the enzymes that would cause decay. Most vegetables are simply blanched *(pages 16-17),* then dry-packed or tray-frozen without sugar.

Certain root vegetables, and mushrooms as well, taste their best when wet-packed *(pages 18-19):* They are briefly cooked in butter or oil, then packed with the butter or oil, which helps preserve moisture and flavor. (In fact, most vegetables can be wet-packed, but blanching or tray-freezing keeps them closest to their natural state.)

Other than the trimming and careful wrapping shown on pages 24-25, fresh meats and fish usually need little treatment before freezing; exceptions are indicated in the guide. Shellfish should be shucked or shelled—after cooking, if noted—and packed in rigid containers.

Because the bacteria that spoil any food cannot grow at freezer temperatures, food will remain safe to eat no matter how long it is kept frozen. However, chemical changes in food are only slowed, not halted, by freezing, and the flavor, texture and appearance of frozen food will gradually deteriorate. The storage-life periods given in the guide indicate how long you can expect the food to remain in good condition. Times for a particular food can vary: If the freezer is not kept very full or if your freezer thermometer indicates that the temperature does not remain at 0° F., use the foods within the shorter time periods listed.

To enjoy foods at their best, label every container or package with its contents and quantity, the date of freezing and the day by which the food should be used. Write the labels with a waterproof pen to prevent fading or smearing. And keep an up-to-date list of your freezer's contents, crossing out each item as you remove it.

Peaches, clingstone. Wet-pack with acidulated syrup. Storage life: 10 to 12 months.

Persimmons. Purée the flesh. Dry-pack with ascorbic acid. Storage life: 10 to 12 months.

Pineapple. Dry-pack with sugar or wet-pack. Storage life: 10 to 12 months.

Plums. Dry-pack with sugar or wet-pack. Storage life: 10 to 12 months.

Quinces. Dry-pack with sugar or wet-pack. Storage life: 10 to 12 months.

Rhubarb. Dry-pack with sugar or tray-freeze. Storage life: 10 to 12 months.

Strawberries. Dry-pack with sugar. Storage life: 10 to 12 months.

Meats

Beef. Storage life: nine to 12 months.
Chicken. Storage life: six to seven months.
Duck. Storage life: six to seven months.
Game animals. Draw, dress, cut into pieces, remove all fat, and refrigerate for 24 to 36 hours—until flesh is no longer rigid—before wrapping and freezing. Storage life: six to nine months.
Game birds. Dress and pluck dry, and refrigerate for 24 hours—until flesh is no longer rigid—before wrapping and freezing. Storage life: six to 12 months, depending on fattiness.
Goose. Storage life: three to four months.
Ground meat. Storage life: four

to six months.
Lamb. Storage life: nine to 12 months.
Pork. Storage life: six to nine months.
Veal. Storage life: four to six months.

Variety Meats

Bone marrow. Storage life: one month.
Giblets. Remove all fat. Storage life: two to three months.
Hearts. See Giblets.
Kidneys. Halve, and remove all fat. Storage life: one to two months.
Liver (beef, lamb, veal). Storage life: three to four months.
Liver, pork. Storage life: one to two months.
Tongue. Remove all fat. Storage life: six weeks.

Fish and Shellfish

Clams. Shuck, dip in brine, wet-pack with natural juices and brine. Leave ½ inch [1 cm.] headspace. Storage life: three to four months.
Crab. Cook, shell, dry-pack meat leaving ½ inch [1 cm.] headspace. Storage life: three to four months.
Fish, fat (eel, herring, mackerel, salmon, trout, whitefish). Storage life: two to three months.
Fish, lean (flounder, haddock, halibut, rockfish, snapper). Storage life: three to six months.
Lobster. See Crab.

Mussels. See Clams.
Oysters. See Clams.
Scallops. See Clams.
Shrimp. Clean but leave in shell. Dip in brine. Dry-pack, leaving no headspace. Storage life: four to six months.

Prepared Foods

Juice, citrus. Pour into rigid containers, leaving 1 inch [2½ cm.] headspace. Storage life: three to four months.
Juice, other fruits. Pour into rigid containers, leaving 1 inch [2½ cm.] headspace. Storage life: seven to eight months.
Juice, vegetable. Pour into rigid containers, leaving 1 inch [2½ cm.] headspace. Storage life: 10 to 12 months.
Nectar, fruit. Pour into rigid containers, leaving 1 inch [2½ cm.] headspace. Storage life: seven to eight months.
Purée, tomato. See Purées, fruit.
Purées, fruit. Pour into rigid containers, leaving 1 inch [2½ cm.] headspace. Storage life: seven to eight months.
Stock, meat. Pour into rigid containers, leaving 1 inch [2½ cm.] headspace. Storage life: 10 to 12 months.
Stock, pectin. Pour into rigid containers, leaving 1 inch [2½ cm.] headspace. Storage life: 10 to 12 months.
Stock, vegetable. Pour into rigid containers, leaving 1 inch [2½ cm.] headspace. Storage life: 10 to 12 months.

How to Blanch Vegetables — and Why

Except for green peppers and onions, whose low enzyme content protects them, vegetables to be frozen require preparations that deactivate their enzymes, which otherwise would cause withering. As indicated in the guide on pages 14-15, most vegetables can be prepared in several different ways, but the usual treatment is blanching, which leaves the food as close to its fresh state as possible.

For blanching, vegetables should be washed and trimmed as they would be for cooking: Cauliflower should be divided into florets, for instance, and spinach should be stemmed. Separate the tender tips from the denser stalks of such vegetables as asparagus: The stalks require longer blanching.

The blanching method you choose depends on the vegetable. Most vegetables—green beans, for example—should be plunged briefly into boiling water. For easy handling, use a colander or wire basket to hold the vegetables.

Delicate vegetables—spinach, asparagus tips and broccoli florets—should be gently steamed over water. The best equipment for this is a steamer with a basket large enough to hold the vegetables in a single layer, thus ensuring that they will cook evenly.

All blanching is brief and the timing must be precise: Insufficient blanching leaves active enzymes; prolonged blanching softens the vegetables. When the blanching is complete, plunge the vegetables into ice water to stop the cooking.

The blanched vegetables are usually dry-packed without added liquid in plastic freezer bags or rigid containers. Many vegetables can be spread on trays to freeze—a process called tray-freezing—then put into containers. Packed this way, the frozen vegetables remain separate, and you can remove whatever quantity you wish. Leaves cannot be kept separate; they are simply sealed in bags.

All frozen vegetables should be cooked for serving—freezing softens them too much to be used raw. Because thawing would collapse their cells, most vegetables—except corn on the cob, whose kernels would cook before the cob thawed—should be cooked right from the freezer. Plunge them into boiling water and cook them for a third of the normal time.

Immersion in Boiling Water

1 **Preparing the beans.** Snap off both ends of each green bean, remove strings if necessary, and place the bean in a large bowl. When all of the beans are trimmed, fill the bowl with cold water and repeatedly push the beans down to clean them. Pour off the water.

2 **Blanching the beans.** Bring a large pot of water to a boil over high heat. Place no more than a handful of beans in a wire basket or colander — a large amount of beans would lower the water temperature and stop boiling — and plunge the colander into the water. Let the beans boil for two minutes.

Quick Exposure to Steam

1 **Trimming spinach.** Fold each spinach leaf in half lengthwise, exposing its stem, and hold its glossy sides firmly together. Pull the stem end up along the length of the leaf to remove it.

2 **Cleaning the leaves.** Immerse the spinach leaves in a bowl of cold water and swish them gently up and down. Lift out the leaves and discard the water and accumulated sand. Refill the bowl and repeat the rinsing operation until no grit appears in the bowl. Drain the leaves in a colander.

3 **Cooling the beans.** Lift the basket of beans from the water and plunge it immediately into a bowl of ice water to arrest the cooking. Let the beans cool in the water for two minutes, then drain them and spread them out on a double layer of paper towels to dry.

4 **Tray-freezing.** When all of the beans have been blanched and dried, arrange them in rows on a flat metal tray; a jelly-roll pan is used here. Leave enough space around each bean to keep it from sticking to its neighbors. Cover the tray tightly with foil and place it in the freezer. The beans should freeze hard in one to two hours.

5 **Sealing in bags.** Remove the beans from the freezer and quickly pack them in plastic freezer bags, allowing only as many beans for each bag as you plan to cook at one time. Press the bags gently to squeeze out excess air, tape them closed with freezer tape, label each bag with the date of freezing, and return the beans to the freezer.

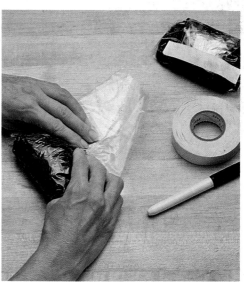

3 **Steaming and cooling.** In a large steamer, bring 1 to 2 inches [2½ to 5 cm.] of water to a boil. Spread a shallow layer of spinach leaves inside the steamer basket. Fit the basket into the steamer, cover, and steam for one and a half minutes. Turn the spinach into a bowl of ice water *(above)*, or immerse the basket in the water, for two minutes.

4 **Drying the leaves.** Spread the cooled spinach leaves on a double layer of paper towels. With a large, doubled sheet of paper towel, gently pat the leaves dry. Do not press hard on the leaves to squeeze out moisture: Rough handling will bruise them.

5 **Packing the spinach.** Divide the leaves into serving portions and pack them into plastic freezer bags. With your fingers, squeeze excess air from the bags. Seal the bags with freezer tape, mark the date of freezing on the tape and freeze the bags at once.

Ready-to-Serve Vegetables

Blanching is the usual preliminary to freezing vegetables *(pages 16-17)*, but any heat checks enzyme action; other cooking methods protect vegetables unsuited to blanching or yield prepared vegetable dishes frozen for convenience.

Of the alternative cooking methods, the most common is to sauté vegetables in butter or oil, then freeze them. That process—like others described here—is known as wet packing, because the vegetables are packed in their pan juices, which help exclude air during freezing. Almost any vegetable can be sautéed, but the method is required for summer squash, tomatoes and mushrooms; they would be damaged by blanching. However, sautéed vegetables have a shorter storage life *(pages 14-15)* than blanched ones: The fat eventually turns rancid.

Sautéing also can be used to fashion root vegetables into foundations for more elaborate dishes *(right, top)*. The vegetables—turnips, in this case, although rutabagas or parsnips could be used instead—are shredded, sautéed in butter and frozen in a block. To turn the frozen turnips into an appetizing gratin, unwrap the block and thaw it in a buttered pan in a 350° F. [180° C.] oven. After about 20 minutes, when the turnips are soft, pour heavy cream over them, sprinkle them with toasted bread crumbs, raise the heat to 400° F. [200° C.] and bake them for 30 minutes, until golden.

Even more homogeneous effects can be obtained by freezing vegetable purées—vegetables boiled until tender, then puréed through a food mill or sieve or in a food processor. Puréeing is a fine treatment for tomatoes. They can be cut up, stewed in their own abundant juice, and sieved to make a sauce that freezes well.

Onions and peppers need not be cooked before freezing because they are low in enzymes. If you like, however, they can be chopped and sautéed, then frozen. Peppers, in addition, lend themselves to a special presentation. When they are broiled, their skins can be removed easily, and their flesh gives off copious juice. In the bottom demonstration, peppers are broiled whole, so that their juice collects inside them; they are then frozen in the juice. Thawed, the peppers can be eaten cold, or warmed and served hot.

Sautéed Turnips for a Savory Gratin

1 **Shredding turnips.** Peel the turnips and cut them into pieces that will fit inside a rotary shredder. Use the medium disk of the shredder to cut the turnips into thin strips *(above)*. Alternatively, shred the vegetables coarse with a box grater.

2 **Sautéing the turnips.** Melt butter in a pan, add the shredded turnips and season them with a little salt. Stirring frequently to ensure even cooking, sauté the vegetables over low heat until slightly soft but not colored—about five minutes. Spread them in a dish to cool.

Roasted Peppers Frozen in Their Juice

1 **Peeling peppers.** Broil whole red and green peppers, turning them occasionally, for four to 10 minutes, until the skins are blistered and charred on all sides. Cover the cooked peppers with a damp towel: Steam will loosen their skins. When the peppers are cool, hold each one over a bowl to catch juice, and pull off the skin *(above, left)*; then pull out the cores and attached seeds *(right)*. Drop the skins and cores into the bowl, and place the peeled peppers in a second bowl.

3 **Packing the turnips.** Line a shallow gratin dish with a piece of plastic wrap slightly more than double the size of the dish. Spoon enough turnips into the dish to fill it to just below the top and fold the film tightly over them. Pack any remaining turnips in dishes. Place the packed dishes in the freezer.

4 **Unmolding the vegetables.** When the sautéed turnips are frozen hard — after several hours — remove the dishes from the freezer. Tip the wrapped turnips out of the dishes *(above)*. If the turnips do not come out easily, loosen them by dipping the bases of the dishes into hot water for a few seconds.

5 **Wrapping the packs.** Place each frozen pack of turnips on a double thickness of heavy-duty foil. Tightly wrap the packs, using the drugstore-wrap technique shown on pages 8-9. Mark each pack with the date of freezing and stack the packs in the freezer.

2 **Removing seeds.** On a plate, cut open each pepper and, using a spoon, scrape out any seeds. Return the cleaned peppers to their bowl. Put the seeds and any juice from the plate into the bowl containing the skins and cores.

3 **Straining the juice.** Set a strainer over the bowl of peppers and tip the skins, cores and seeds into it so that the juice drains into the bowl beneath. Discard the debris from the strainer.

4 **Packing the peppers.** Place the peppers in rigid freezer containers. Pour the juice from the bowl over the peppers. Put the lids on the containers, mark them with the date of freezing, and freeze the peppers.

Simple Strategies for Fruits

Because of their acidity *(page 13),* most fruits are safe from deterioration caused by enzymes. However, they require a few simple preliminary treatments—varying in many cases with the type of fruit—before they can be frozen successfully.

One universal rule is that the fruit must be at its peak—ripe but firm. Remove peels, stems, cores or pits before freezing and, depending on their future use, cut up large fruits. This will minimize handling of the fruit when it is thawed and fragile.

Fruits that contain sturdy, moisture-retentive cells—blueberries and raspberries, for instance—can be tray-frozen *(far right).* But most fruits render their juice as they freeze and, without support, their cell walls would collapse and their texture be ruined. These fruits must be frozen in sugar; as they lose their liquid, they will absorb the sugar and remain plump and firm.

Dark-colored and juicy fruits—cherries, strawberries and pineapple, for example—can be dry-packed with sugar, as in the top demonstration here. In freezing, the fruit juices will dissolve the sugar to a syrupy glaze.

Chemical compounds called tannins darken the flesh of pale fruits—apples, pears, apricots and peaches—when exposed to air. In order to keep these fruits from discoloring, wet-pack them in syrup, made by boiling sugar and water together *(recipe, page 165),* using 1 to 2 cups [250 to 500 ml.] per quart [1 liter] of fruit. As an added defense, acidulate the syrup with ascorbic acid or lemon juice *(pages 14-15).*

Always leave an inch [2½ cm.] of space at the top of the container to allow for the liquid's expansion during freezing. To keep buoyant fruit submerged, top it with a piece of crumpled wax paper before covering the container.

Fruits can also be frozen as purées—either sweetened or plain, cooked or raw. Purées can be made from bruised fruit, although the damaged parts should be cut away. If you purée pale fruits, add ascorbic acid before freezing.

To thaw frozen fruit for eating or for use in cooking, leave it in its wrapping in the refrigerator. When thawed, its texture will be softer than that of fresh fruit.

Coating Slices with Sugar

1 **Peeling a pineapple.** With a long, sharp knife, cut off the top and trim the base of the fruit. Stand the pineapple upright and slice off the rind in vertical strips. Use the tip of a small knife to remove the eyes, or spikes of skin, that remain embedded in the flesh.

2 **Slicing.** Lay the pineapple on its side. Steady the fruit with one hand and use the long knife to slice through the pineapple at ¾-inch [1½-cm.] intervals.

Making a Smooth Purée

1 **Preparing stoned fruit.** Wash and dry the fruit—in this case, apricots—and pull off any stems. To remove its pit, cut around each apricot lengthwise, slicing down to the pit; hold the apricot with both hands and gently twist the halves in opposite directions to pull them apart *(above).* Lift out the exposed pit and discard it.

2 **Puréeing the fruit.** If you use a food mill, as here, set it over a bowl and purée the apricots a handful at a time. After each batch, discard the skins from the food mill. To purée the fruit in a blender or processor, remove the skins first: Dip the apricots in boiling water, drain them, and peel off the loosened skins with a small knife.

Keeping Berries in Shape

3 **Removing the core.** To make neat pineapple rings, use a small, round cookie cutter to remove the rough, fibrous core from the center of each slice *(above)*. Or, cut out the cores with a small, sharp knife. Discard the cores.

4 **Sugaring.** Sprinkle a shallow layer of sugar into a rigid, round container that will hold the slices snugly. Stack pineapple slices in the container, sprinkling each slice lightly with more sugar. Fill the container to within ½ inch [1 cm.] of the top, ending with a layer of sugar. Cover and freeze.

1 **Tray-freezing berries.** Pick over the fruit — in this case, raspberries — and pull off any stems. Arrange the berries in a single layer on a tray, leaving space between them so that they do not freeze in clumps. If you like, sprinkle the berries with sugar as extra insurance against softening. Cover the tray and place it in the freezer.

3 **Packing.** Stir about 1 cup [¼ liter] of sugar and 3 tablespoons [45 ml.] of lemon juice into every 4 cups [1 liter] of purée. To freeze the purée in neat blocks, ladle it into plastic freezer bags set inside rigid containers *(above)*. Leave a 1-inch [2½-cm.] space at the top of the containers. Press the bags to expel air, twist the tops and tie them closed.

4 **Freezing the purée.** Put the lids on the containers and place them in the freezer. When the purée is frozen solid — after three or four hours — remove the containers from the freezer, take off the lids and tip out the blocks. Stack the blocks of frozen purée in the freezer.

2 **Packing the fruit.** When the berries are frozen — after one or two hours — transfer them to plastic freezer bags or to rigid containers. To seal a bag, squeeze it with your hands to expel the air, twist the bag at the neck and tie it. Return the fruit to the freezer.

Convenient Packaging of Poultry

As with vegetables and fruits, the key to successful freezing of poultry is to use fresh, carefully prepared ingredients.

Before freezing, any bird should be thoroughly cleaned. Pluck out or singe off any pinfeathers. Take out the giblets (heart, liver and gizzard), wash and dry them, and freeze them separately; giblets have a shorter storage life than the rest of the bird. Rinse the bird inside and out with cold water and dry it.

Birds to be frozen whole should never be stuffed. Stuffings usually include onions, herbs or spices, which freeze less well than poultry does and may develop a tainted flavor while the bird is still in good condition. Stuffing ingredients also thaw at a different rate from the meat, and the varying temperatures within a thawing, stuffed bird would encourage the growth of bacteria.

Birds frozen whole should be trussed beforehand. Trussing yields a compact shape that can save space in the freezer, and makes it possible to roast the bird without further preparation as soon as it has thawed. The method shown below is especially convenient: The bird is held in shape by a single string, which is easy to remove if you wish to stuff the thawed bird before cooking it. Although a chicken is trussed here, the method is equally suited to turkey, duck, goose or game birds such as pheasant. Smaller birds—Rock Cornish game hens or quail, for instance—need only be tightly enclosed in freezer paper to hold their shapes.

If you intend to cook pieces of a bird separately, cut it up before you freeze it. Divided in the manner demonstrated at right, a chicken will yield five pieces of approximately equal size. The breast of a large bird—a capon, goose or turkey, for example—can be further divided into two or more portions. The trimmings—the bird's back and wing tips—can be frozen on their own and used for stock.

When you pack a cut-up bird for the freezer, separate the pieces with plastic freezer wrap, freezer paper or heavy-duty foil. Individual pieces can then be used without thawing the whole parcel.

Once frozen, lean birds such as chickens and turkeys will stay in good condition longer than birds such as ducks and geese, whose high proportion of fat may turn rancid. The guide on pages 14-15 gives recommended storage times.

Before cooking, thaw birds in loosened wrappings in the refrigerator, allowing two hours for each pound [½ kg.] of meat. Unless completely thawed, the cut-up parts or whole bird will not cook evenly: The surface will cook before enough heat penetrates the meat to kill any bacteria that are present.

Cutting Serving Portions

1 **Cutting along the spine.** Lay the bird—in this case, a chicken—on its breast. Make a shallow cut across the back below the shoulder blades and a shallow cut along the spine from the center of the first cut to the tail.

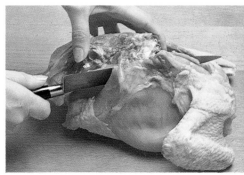

5 **Removing the back.** Insert the knife in the cavity, cutting edge up. Pierce one side between the shoulder joint and rib cage, and slit through the ribs parallel to the spine. Repeat on the other side (above) and pull away the back.

Trussing to Save Space

1 **Covering the neck cavity.** Remove the giblets, and rinse and dry the bird—a chicken, here. Place the bird breast side down. Fold the loose flap of skin at the neck over the cavity.

2 **Positioning the wings.** Fold the wing tips of the bird up and over the shoulder joints (above); then turn the bird on its back. Cut a piece of string about 3 feet [1 m.] long.

3 **Trussing drumsticks.** Center the string under the bird's tail. Loop each end over and around the opposite drumstick. Pull on the ends to draw the drumsticks and tail over the vent.

2 **Freeing the oysters.** The oysters are two morsels of flesh that lie on either side of the spine below the shoulder blades. Using the point of a knife and your fingers, free the oysters from the bone; leave them attached to the skin.

3 **Removing the legs.** Set the bird on its back. Cut the skin between the thighs and body, and bend the legs to pop out the thighbones. Remove the legs with the oysters attached. Cut the bony knots from the drumsticks.

4 **Freeing the shoulders.** Turn the bird on its breast. Insert the knife between the spine and each shoulder blade, and cut down firmly into the cavity. Do not cut through the bird: Leave the wings attached to the breast.

6 **Removing the wings.** Turn the breast and wings skin side up. To sever the wings, cut through the skin and flesh from where the collarbone meets the breastbone. Cut diagonally to include breast meat with each wing.

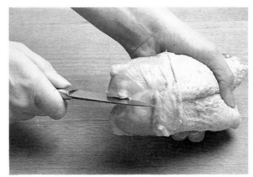

7 **Cutting the tendons.** In order to prevent the legs from becoming misshapen while they are cooking, bend each drumstick back to expose the strong tendons between the drumstick and the thigh. Slice through the tendons.

8 **Wrapping.** Arrange the pieces on plastic freezer wrap in one overlapping layer. Tuck a square of wrap around each piece in order to keep it separate. Close the package. Then wrap it in heavy-duty foil, label and freeze.

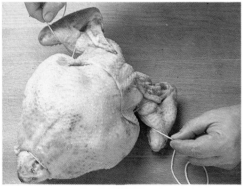

4 **Securing the thighs.** Turn the bird on its breast. Bring the string ends across the back of the thighs, loop the ends around the wings and pull the strings firmly across the neck flap.

5 **Tying the string.** Tie the two ends of string securely in a knot at the neck end of the bird (above). Using scissors or a sharp knife, trim off any excess string.

6 **Wrapping the bird.** Enclose the bird in plastic freezer wrap, pressing out as much air as possible. Cover the parcel with heavy-duty foil, label the parcel, and place it in the freezer.

A Range of Meat-handling Techniques

Like all foods, meat must be frozen quickly to minimize the size of ice crystals that form within it and soften its texture. It is best, therefore, to cut meat into small pieces, which freeze through rapidly.

The precise way you prepare meat for the freezer will depend on how you want to use it later. Cut stew meat into cubes *(right, top);* interleave meat patties and chops with strips of freezer paper *(opposite, top)* for easy separation and even thawing; scoop the marrow from bones *(Step 1, right, below)* before freezing it.

To prevent meat from drying out during storage, wrap it in plastic freezer wrap or freezer paper, or pack ground meat and meat stock in plastic freezer bags or rigid containers. For parcels containing steaks or chops with protruding bones, use an extra wrapping of aluminum foil to protect against punctures.

Since bacteria cannot multiply when frozen, meat remains safe to eat indefinitely. But if meat is kept too long, its quality will deteriorate because of enzyme activity. Enzymes—and the air trapped with wrapped meats—also will eventually turn fat rancid. The rate at which fats turn rancid, however, depends primarily upon their precise chemical composition—whether they are saturated or unsaturated.

Beef and lamb contain a high proportion of saturated fat and thus have a long storage life. The unsaturated fats found in pork react readily with oxygen, and become rancid much more quickly. Cured meats such as bacon contain salt, which encourages oxidation and rancidity; these meats have the shortest storage lives of all. For good results, freeze meats no longer than the periods recommended in the guide on pages 14-15.

Thaw frozen meat before cooking to ensure that it cooks through evenly. Small, thin cuts, such as steaks or chops, may easily overcook if they go into the pan or broiler still frozen. If large pieces are incompletely thawed, their centers can remain uncooked and may possibly be dangerous, because meat is more vulnerable to bacterial growth after freezing. Loosen the wrapping and thaw the meat slowly in a refrigerator, allowing about five hours per pound [½ kg.] of meat. Once thawed, cook it as soon as possible.

Cubing Large Cuts for Stews

1 **Trimming the meat.** With a sharp knife, cut excess fat from the meat — in this instance, beef chuck roast. Divide the meat along the muscle seams into lean sections. Trim away any tough connective tissue that remains attached to the meat *(above).*

2 **Wrapping the meat.** Cut the trimmed meat into conveniently sized chunks — here, pieces about 2 inches [5 cm.] square. Enclose the meat — in one layer for even freezing — in plastic freezer wrap, in recipe-sized portions. Wrap each package in foil, using the drugstore wrap shown on pages 8-9. Label the package and freeze it.

Extracting Marrow from Beef Bones

1 **Extracting marrow.** Ask your butcher to saw beef shank bones into 3- to 4-inch [8- to 12-cm.] sections. This will make it easier to remove the marrow. Loosen the marrow from one end of the bone with a small, sharp knife, and gently prize it out until the bone is empty.

2 **Wrapping the marrow.** Enclose the marrow in one layer in plastic freezer wrap, using a drugstore wrap. Press down on the package with your hands to exclude air *(above).* Then wrap the parcel in heavy-duty foil. Label the package and freeze it.

Shaping and Wrapping Patties

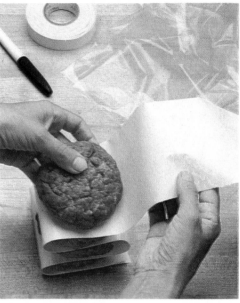

1 **Chopping the meat.** Trim all fat and connective tissue from the meat — top-round roast is shown. Slice the meat into small, even-sized cubes and bunch them in a single layer on a work surface. Using two knives of equal weight and a rhythmic, drum-beating motion, chop the meat to the texture you desire.

2 **Forming the patties.** Cut the layer of chopped meat into equal-sized portions. Gather each portion into a ball, then press it firmly into a thick patty. Cut freezer paper into a long strip that is double the width of the patties, and fold the paper in half lengthwise, dull sides in.

3 **Wrapping the meat.** Set a patty on one end of the paper. Fold the strip up over the patty to cover it. Set a second patty on top of the paper, centering it over the first. Repeat this procedure with the rest of the patties. Then put the stack in a plastic freezer bag, press out air, tape the bag closed, label and freeze.

Dealing with Protruding Bones

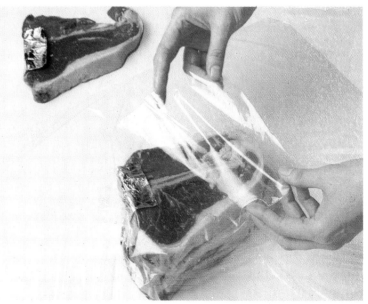

1 **Padding bones.** With a sharp knife, trim excess fat from the meat — in this case, pin-bone sirloin steaks. To prevent the meat bones from piercing and tearing their package and thus exposing the meat to the air, fold small pieces of foil into pads and wrap the pads over the bones (above).

2 **Separating steaks.** Place a steak on a long sheet of plastic freezer wrap. Cover the meat with a square of plastic wrap; set the next steak on top. Repeat with the remaining meat, separating each steak from the next. Bring the ends of the bottom piece of freezer wrap over the steaks to cover them tightly. Enclose the parcel in heavy-duty foil, using a drugstore wrap, label it and freeze it.

Special Care for Fish

Fish are the most perishable of foods, requiring special freezing procedures to prevent discoloration and rancidity. The first consideration is that the fish be impeccably fresh, preferably just hours from the water. Clean and gut them as soon as possible; the viscera contain powerful digestive enzymes that make the flesh spoil quickly. Rinse away every trace of blood; otherwise, the flesh will discolor during freezing.

To minimize handling of the fish when you thaw them for cooking, freeze them the way you plan to use them. Small fish such as trout or striped bass usually are frozen whole *(top demonstration)*. Larger fish such as salmon, swordfish or cod may be cut into steaks or fillets *(bottom demonstrations)*. However you prepare the fish, dip them for 30 seconds into a salt-and-water brine. The brine firms the flesh by drawing out moisture, thereby reducing the amount of leakage when the fish are thawed; it also helps prevent discoloration.

You may freeze the prepared fish conventionally by enclosing them in plastic freezer wrap. Fillets or steaks should be interleaved with the wrap to keep them separate. Overwrap the packages in heavy-duty foil or freezer paper.

An extra step—glazing—will ensure the best possible protection of the delicate flesh. First, freeze the fish on a loosely covered tray until the flesh has frozen solid. Then dip the fish into ice water; when the water comes in contact with the frozen fish, it will harden instantly into a thin coating of ice. Repeat the dipping several times to build up an airtight glaze. Finally, enclose the fish in heavy-duty foil or freezer paper for freezing.

Glazed lean fish such as cod, lake perch or flounder will keep for up to six months in the freezer, and oily fish—with their greater tendency toward rancidity—will keep for up to three months. For optimal results, add an extra glazing of ice every two months.

To thaw fish, loosen their wrappings and transfer them to the refrigerator. Allow three to eight hours per pound [½ kg.] of fish for defrosting, and cook the fish as soon as they have thawed.

Applying an Icy Glaze

1 **Gutting the fish.** Scrape the scales from a whole fish—here, a striped bass. Slice off the pectoral fins on either side of the head. Slit the belly open from just below the gills down to the vent. Pull out all of the entrails and discard them.

2 **Completing the cleaning.** Rinse the cavity of the fish under cold running water. Run your thumbs or a sharp knife down either side of the backbone inside the fish to release blood pockets. Pull out the gills. Thoroughly rinse the fish again.

How to Cut Steaks

1 **Beginning the cut.** As shown in Steps 1 and 2 *(top)*, clean and gut a large—at least 10-pound [5-kg.]—fish; salmon is used here. Cut off the head just behind the gills. Lay the fish on its belly and, with a large, sharp knife, cut across the fish down through the backbone about 1½ inches [4 cm.] behind the first cut.

2 **Finishing the steak.** Turn the fish on its side and cut down through the body to sever the steak. Cut the rest of the steaks in the same way. Dip the steaks in brine and, if you wish, glaze them *(Steps 3 and 4, opposite top)*. Wrap the steaks in freezer paper, interleaving them *(page 25)* so they will stay separate, and freeze them.

3 **Dipping in brine.** Stir 1 cup [¼ liter] of salt into 1 gallon [4 liters] of water until the salt dissolves. Dip the fish in this brine, keeping it immersed for 30 seconds. Drain the fish on paper towels, then lay it on a baking sheet, cover it loosely with foil, and freeze.

4 **Glazing the fish.** When the fish has frozen solid — after about four hours — remove it from the freezer. Immerse the fish in a shallow pan of ice water for about five seconds *(above, left)*. Lift the fish and hold it in the air for a few seconds *(right)* to allow an ice coat to form. Repeat two or three times. Enclose the fish in freezer paper, using a drugstore wrap *(pages 8-9)*, and freeze it.

Slicing and Skinning Fillets

1 **Cutting the first fillet.** Slice the fins from a whole, cleaned fish — in this instance, a red snapper. Slice across the body just behind the gills, almost severing the head. Slice open the back, exposing the backbone. Lift the fillet and insert the knife between the flesh and ribs, parallel to the ribs. With short strokes, cut the fillet from the ribs.

2 **Finishing the fillet.** Insert the knife parallel to and under the backbone at the head of the fish. With short strokes, cut down the length of the fish, lifting the backbone to free the second fillet. Discard the skeleton or freeze it in a plastic freezer bag for use in fish stock.

3 **Skinning.** Lay each fillet skin side down. Slice about ½ inch [1 cm.] of flesh away from the skin at the tail end. Insert the knife at a slight angle between skin and flesh, and pull the skin toward you to separate skin and flesh. Dip the fillets in brine and, if you like, glaze them *(Steps 3 and 4, top)*. Pack them in freezer paper and freeze.

2
Canning
Heat-imposed Protection

"Canning" is something of a misnomer: It generally means packing foods into jars rather than cans, which are not readily available, and then heating the jars. The heat kills spoilage organisms in the food and creates a partial vacuum that seals the jars and prevents contamination. Such heat processing is not difficult and is rewarding: The preserves retain their quality and flavor for long periods without refrigeration.

Three types of spoilage organisms are present in all food: enzymes; molds and yeasts, which cause fuzzy growths and unwanted fermentation; and bacterial spores, some of which become deadly poisons if allowed to grow. Two heat-processing methods are used to check these spoilage agents, the choice depending on the food involved. Fruits, for instance, contain acid, which inactivates enzymes and potentially poisonous bacteria. The acid does not harm molds or yeasts, but such organisms are destroyed by temperatures ranging from 140° to 190° F. [60° to 88° C.]. Acid foods are therefore safely preserved when they are heat-processed in boiling water *(pages 32-33)*, whose temperature reaches 212° F. [100° C.]. Low-acid foods—vegetables other than tomatoes, and all meats and poultry—have no natural protection against enzymes and bacteria, and higher heat is required to destroy these spoilage agents. Low-acid foods must be processed in special pressure canners *(pages 40-41)*, which raise the temperature of boiling water to at least 240° F. [116° C.], the point at which enzymes are inactivated and bacteria die.

The length of time food must spend either in a boiling water bath or a pressure canner depends on its density, its temperature when processing begins, and the size of jar in which it is packed. Food canned in large pieces will require longer processing to heat it through than food that is cut small. Food packed raw and cool in jars *(pages 32-35)* takes longer to reach the necessary temperature than food that is cooked and packed hot *(pages 36-37)*. Food packed in large jars needs longer processing than food packed in small jars. The guide on pages 30-31 lists processing methods and times for 65 frequently canned foods.

To protect its color and flavor, canned food should be stored in a spot that is dark and cool—40° to 60° F. [5° to 15° C.]. Most foods will keep for a year, but any food whose odor or appearance is questionable when the jar is opened should be discarded at once. Never test it by tasting it.

A syrup made with red wine and fragrant with spices is ladled into a jar of pears through a wide-necked preserving funnel. The pears were poached in the syrup until tender, then packed closely into the jars. After the jars have been closed, processing in a boiling water bath will sterilize their contents.

A Primer for Home Canning

The guide below specifies preliminary preparations, as well as processing methods and times, for canning fruits, vegetables, meats and a range of prepared food. The foods included on the list are those whose flavor, color and texture remain closest to their fresh state when they are canned—and those that may be canned with absolute safety. (A few foods were excluded because it is unsafe to can them at home; pumpkin, which is too dense for home canning, is one of these.)

Before canning, foods should be prepared—cleaned, peeled, cored or seeded—as they are for ordinary cooking. The pieces of food you pack in canning jars must be of uniform size to ensure even heat penetration. Some foods require additional treatment, and these are noted in the guide. Pale-fleshed fruits such as peaches, for instance, should be acidulated after peeling to prevent their

exposed flesh from darkening; to do this, drop the fruits into water that has been mixed with lemon juice, allowing 1 teaspoon [5 ml.] of juice per quart [1 liter] of water. Almost all vegetables should be water-blanched—cooked briefly in boiling water (pages 16-17)—to soften them slightly before canning; the blanching time for each vegetable is in the guide.

Prepared foods may be packed into canning jars in two basic ways, demonstrated on the following pages. Some foods—notably fruits and vegetables—are raw-packed, a procedure that consists of simply placing uncooked or briefly blanched food in canning jars and covering them with boiling liquid (pages 34-35). Other foods are cooked before canning and transferred from the cooking pot to the jars, an approach known as hot packing (pages 36-37). Though certain foods may be packed by either meth-

od, the guide lists only the one that results in the best flavor, texture and color.

No matter which method is used, the food must be packed loosely, with some room—called headspace—left at the top of each jar to allow for expansion of the contents during processing. In most cases, ½ inch [1 cm.] of headspace is adequate. Certain foods—starchy vegetables, for example—require extra headspace; such cases are noted in the guide.

Use only those canning jars described on pages 10-11 and prepared as shown there. The jar sizes specified in the guide are maximum; do not use larger jars. Most food may be canned either in pint [½-liter] or quart [1-liter] jars, but a few foods are so dense that for adequate heating they must be canned in small quantities: The jars permitted may be no larger than a pint or even a half pint [¼ liter].

The processing method used to steril-

A Guide to Preparations and Processing

Vegetables

Artichoke bottoms. Water-blanch for five minutes in a brine composed of 1 gallon [4 liters] of water, ¾ cup [175 ml.] of vinegar and 3 tablespoons [45 ml.] of salt; hot-pack in boiling brine. Process in a pressure canner: quarts, 25 minutes.

Beets. Water-blanch for 15 to 25 minutes to remove skins; raw-pack in boiling water. Process in a pressure canner: pints, 40 minutes; quarts, 50 minutes.

Beans, broad and lima. Water-blanch for three minutes; raw-pack in boiling water, leaving 1 inch [2½ cm.] headspace. Process in a pressure canner: pints, 40 minutes; quarts, 50 minutes.

Beans, green and wax. Do not blanch; raw-pack in boiling water. Process in a pressure canner: pints, 20 minutes; quarts, 25 minutes.

Carrots. Water-blanch for three minutes; raw-pack in boiling water. Process in a pressure canner: pints, 25 minutes; quarts, 30 minutes.

Corn kernels. Water-blanch for three minutes on cob, then cut from cob; raw-pack in boiling water, leaving 1 inch [2½ cm.] headspace. Process in a pressure canner: pints, 55 minutes; quarts, 85 minutes.

Corn on the cob. Small cobs only. See Corn kernels.

Peas, sweet. Water-blanch small peas for three minutes, medium-sized peas for five minutes; raw-pack in boiling water, leaving 1 inch [2½ cm.] headspace. Process in a pressure canner: pints and quarts, 40 minutes.

Potatoes, sweet. Do not blanch or purée. Raw-pack with boiling water or sugar syrup, leaving 1 inch [2½ cm.] headspace. Process in a pressure canner: pints, 55 minutes.

Tomatoes. Raw-pack in own juice, adding 1 tablespoon [15 ml.] of bottled lemon juice per pint. Process in a boiling water bath: pints, 35 minutes; quarts, 45 minutes.

Fruits

Apples. Acidulate. Simmer for five minutes in sugar syrup; hot-pack in syrup. Process in a boiling water bath: pints, 15 minutes; quarts, 20 minutes.

Apricots. Acidulate. Raw-pack in sugar syrup. Process in a boiling water bath: pints, 25 minutes; quarts, 30 minutes.

Blueberries. Simmer for 30 seconds in water; cold-pack in syrup. Process in a boiling water bath: pints, 15 minutes; quarts, 20 minutes.

Cherries. Raw-pack in sugar syrup. Process in a boiling water bath: pints, 20 minutes; quarts, 25 minutes.

Cranberries. See Blueberries.

Currants. See Blueberries.

Elderberries. See Blueberries.

Gooseberries. See Blueberries.

Grapefruit. Raw-pack in sugar syrup. Process in a boiling water bath: pints or quarts, 10 minutes.

Grapes. Raw-pack in sugar syrup. Process in a boiling water bath: pints, 15 minutes; quarts, 20 minutes.

Nectarines. See Apricots.

Oranges. See Grapefruit.

Peaches. See Apricots.

Pears. Acidulate. Simmer for 20 minutes in sugar syrup; hot-pack in syrup. Process in a boiling water bath: pints, 15 minutes; quarts, 20 minutes.

Pineapple. Simmer for five minutes in sugar syrup; hot-pack in syrup. Process in a boiling water bath: pints, 15 minutes; quarts, 20 minutes.

Plums. Raw-pack in sugar syrup. Process in a boiling water bath: pints, 20 minutes; quarts, 25 minutes.

Rhubarb. Steep cut-up stalks with sugar for four hours. Simmer for one minute in its own juice; hot-pack in juice. Process in a

ize the food and seal the jars depends on the food involved. Acid foods or those canned with acidic ingredients may be processed in a boiling water bath *(page 33)*. If you use screw-top canning jars, bring the water in the canning vessel to a simmer and place the filled jars in it, making sure the jars are covered by at least 1 inch [2½ cm.] of water. Bring the water to a boil, cover the canner, and begin counting processing time. Remove the jars as soon as processing time is up. If you are using more fragile clamp-top jars, the water in the canner must be tepid to start; bring the water to a boil before beginning to count processing time. At the end of that period, turn off the heat and let the jars cool in the water to room temperature before removing them.

Foods low in acid—meat and vegetables—are processed in a pressure canner set at 10 pounds per square inch [70 kPa, or kiloPascals]; this figure represents the increase over normal atmospheric pressure, which is 14.7 pounds per square inch [100 kPa] at sea level. Only sturdy screw-top canning jars may be used.

Processing times for either a water bath or pressure canner will vary according to the acidity and density of the food and the size of the jars used. For effective preservation, precisely follow the processing times that are specified in the guide. At altitudes more than 1,000 feet [305 m.] above sea level, the boiling temperature of water decreases, and for safe preservation of food you must increase processing time in the boiling water bath and increase pressure in the pressure canner; follow the instructions in the box at right to make these adjustments.

Check each processed jar's seal after it has cooled *(pages 10-11):* An improper seal will allow the food to spoil.

Adjustments for Altitude

When processing at high altitudes, use the following formulas:
Boiling water bath: If the specified processing time is 20 minutes or less, add one minute for each 1,000 feet [305 m.] of altitude above sea level. For longer processing times, add two minutes for each 1,000 feet.
Pressure canner: At sea level, all food is processed at 10 pounds of pressure per square inch [70 kPa]. If your pressure canner has a dial gauge, increase the pressure by ½ pound [3½ kPa] for each 1,000 feet of altitude. For a pressure canner with weights, process all food at 15 pounds [105 kPa] at an altitude of more than 1,000 feet.

---

boiling water bath: pints and quarts, 10 minutes.

Meats
Beef. Braise 1-inch [2½-cm.] cubes in meat stock *(recipe, page 167)* for 15 minutes, until tender; hot-pack in degreased stock, leaving 1 inch [2½ cm.] headspace. Process in a pressure canner: pints, 75 minutes.
Chicken (breasts, drumsticks, thighs). Poach in hot water for 15 minutes, until tender. Hot-pack in cooking liquid, placing skinned breasts in the centers of jars, other pieces skin-side out next to the glass, and leaving 1 inch [2½ cm.] headspace. Process in a pressure canner: pints, 65 minutes; quarts, 75 minutes.
Duck (breasts, drumsticks, thighs). See Chicken.
Game birds (breasts, drumsticks, thighs). See Chicken.
Goose (breasts, drumsticks, thighs). See Chicken.
Lamb. See Beef.
Pork. See Beef.
Rabbit (breasts, drumsticks, thighs). See Chicken.
Turkey (breasts, drumsticks, thighs). See Chicken.
Venison. See Beef.

Prepared Foods
Apple juice. Hot-pack. Process in a boiling water bath: pints and quarts, 30 minutes.
Apple nectar. Hot-pack. Process in a boiling water bath: pints, 15 minutes.
Applesauce. Hot-pack. Process in a boiling water bath: pints, 10 minutes.
Apricot nectar. See Apple nectar.
Berry juices. See Apple juice.
Cherry juice. See Apple juice.
Chicken stock. Hot-pack. Process in a pressure canner: pints, 20 minutes; quarts, 25 minutes.
Cranberry juice. Hot-pack. Process in a boiling water bath: pints and quarts, 10 minutes.
Chutneys. Hot-pack. Process in a boiling water bath: pints and quarts, 10 minutes.
Currant juice. See Apple juice.
Fruit butters. Hot-pack. Process in a boiling water bath: pints, 15 minutes.
Fruit cheeses. Hot-pack. Process in a boiling water bath: pints, 15 minutes.
Fruit jams, marmalades and conserves. Hot-pack. Process in a boiling water bath: pints, 15 minutes.
Fruit jellies. Hot-pack in sterilized jars; seal *(pages 10-11).* Processing is unnecessary.
Fruit pickles. Hot-pack. Process in a boiling water bath: pints and quarts, 15 minutes.
Fruit purées. Hot-pack. Process in a boiling water bath: half pints, 10 minutes.
Ketchups. Hot-pack. Process in a boiling water bath: pints and quarts, 10 minutes.
Meat stock. See Chicken stock.
Mincemeat. Hot-pack. Process in a pressure canner: pints, 20 minutes.
Nectarine nectar. See Apple nectar.
Pectin stock. Hot-pack. Process in a boiling water bath: pints, five minutes.
Peach nectar. See Apple nectar.
Relishes. Hot-pack. Process in a boiling water bath: pints, 15 minutes.
Tomato juice. Hot-pack, adding 1 tablespoon [15 ml.] of bottled lemon juice per pint. Process in a boiling water bath: pints and quarts, 35 minutes.
Tomato paste. Hot-pack. Process in a boiling water bath: half pints, 35 minutes.
Tomato sauce. Hot-pack. Process in a boiling water bath: pints, 30 minutes.
Sauerkraut. Hot-pack. Process in a boiling water bath: pints, 15 minutes; quarts, 20 minutes.
Vegetable pickles. Hot-pack. Process in a boiling water bath: pints and quarts, 15 minutes.

A Raw-packed Favorite

Vine-ripened tomatoes, packed into jars at the height of the summer, will lend garden-fresh flavor and crimson color to stews, soups and sauces throughout the year. The best way to preserve tomatoes in their natural state is by the method called raw packing, shown here. The tomatoes are simply blanched, peeled, put in jars and squeezed to release enough juice to cover them. The jars are then covered and processed *(pages 30-31)*.

To preserve their shape, the tomatoes must be of a size that will fit easily into the jars—and ripe enough so that they yield ample juice when pressed gently. To enhance their flavor, the tomatoes can be salted—about 1 teaspoon [5 ml.] for each quart [1-liter] jar. Many cooks add fresh herbs. Basil leaves are used here; oregano, dill, or thyme is also good.

Because the tomatoes are acidic *(page 29)*, they are processed in a boiling water bath. However, tomatoes vary in acidity. To ensure that all growth of harmful bacteria stops, add 2 tablespoons [30 ml.] of lemon juice to each quart jar. To ensure a proper percentage of acid, the lemon juice must be bottled, not fresh.

Once the jars are filled and closed, they can be loaded onto the rack of a water-bath canner for processing. Alternatively, the jars can be wrapped in cloth towels to prevent breakage and set on a rack in any pot that is at least 2 inches [5 cm.] taller than the jars.

When tomatoes are raw-packed, both the jars and their contents are cool. To prevent jars from cracking when they are lowered into the canner or pot, the water should be at a slow simmer. Start counting the processing time when the water reaches a full boil. After processing, the hot jars should be set on doubled kitchen towels to insulate their bases from cold; place the jars well apart to allow air circulation. Let the jars cool for at least 12 hours before testing the lids for a tight seal *(pages 10-11)*.

1 **Coring tomatoes.** Insert the tip of a small knife just outside the stem end of each tomato. Cut around the stem to remove a conical plug about 1 inch [2½ cm.] deep, lifting the plug out with the knife tip. Turn over the tomato and slit a cross through its skin.

4 **Flavoring with basil.** Place three or four basil leaves in the jar *(left)*. Add salt and lemon juice. Continue packing and pressing tomatoes, leaving a ½-inch [1-cm.] headspace *(above)*. Run a nonmetallic spatula around the inside of the jar to release air bubbles, and cap the jar for processing.

2 **Blanching and peeling.** Using a wire basket or colander, immerse four or five tomatoes at a time in a pot of boiling water for 30 seconds, or until the skins begin to wrinkle. Plunge the tomatoes into cold water to stop the cooking, then transfer them to a bowl. Peel each tomato, a quarter at a time, by grasping the slit corners of skin between a knife blade and your thumb, and pulling the skin toward the stem end to free it.

3 **Packing the jars.** Wash canning jars *(pages 10-11)* and set them upside down on a towel. Assemble fresh basil leaves, salt and lemon juice. Place a few tomatoes in a jar and lightly press against them with the back of a wooden spoon to squeeze out some of their juice. Add a few more tomatoes — filling the jar halfway — and again press out the juice.

5 **Loading the canner.** Fill a water-bath canner halfway with water and bring it just to a simmer over high heat. Set the jars in the canner's rack, lift the rack by its wire handles and lower it into the water, folding the tops of the handles over the jars. Be sure the water covers the jars by at least 1 inch [2½ cm.].

6 **Processing.** Cover the canner, bring the water to a boil and process the tomatoes. Uncover the canner, lift the rack and rest it on the rim of the canner. With tongs, move the jars to a towel-covered counter, standing them upright. Leave the jars undisturbed for at least 12 hours.

Packing Uncooked Fruit in Sugar Syrup

Unlike tomatoes *(pages 32-33)*, fruits do not yield their juice unless you cook, crush or purée them *(page 20)*, thereby sacrificing their texture. If you want the fruits to remain as close to a fresh state as possible, cover them with another liquid when you pack them for canning.

The fruits—plums, pears and peaches are the favorites—may be packed whole, halved or cut into sections *(guide, pages 30-31)*. If you wish to leave the skin on whole fruit, pierce it before packing *(box, opposite)* to prevent bursting. Sectioned fruit should be seeded or pitted, and all of the fibrous skin at the center should be scraped away: The fibers would toughen and discolor during storage.

If you section and pit fruit, you can impart extra flavor to the preserve by cracking open the pits' tough outer shells, peeling the rich-tasting kernels *(Step 3)* and adding them to the packed canning jars. Do not use more than two kernels per jar, however: Kernels contain substances that may be toxic in high concentrations.

Numerous liquids *(recipes, pages 148-151)* may be used to cover the fruit. The usual choice is a simple syrup made by boiling water and sugar together until the sugar dissolves. The syrup prevents the fruit's juices from leaking out and permeates the fruit to a certain extent, plumping and firming it. You can use a light, medium or heavy syrup, according to taste *(recipe, page 165)*; if you like honey, substitute it for half of the sugar.

You can also replace all or part of the water with juice made from fruit that you have set aside from the batch being preserved, or with any other juice that will not overpower the preserved fruit's own flavor. In any case, you will need 1 to 2 cups [250 to 500 ml.] of liquid per quart [1 liter] of fruit.

For a less sweet preserve, the fruit may be covered with fruit juice alone; lacking added sugar, however, the fruit's taste and texture will suffer. Do not use plain water as a liquid: It will leach out the fruit's natural sugar and flavor.

You need not add acid to the syrup or fruit as you would to tomatoes. The fruit itself contains enough acid for preservation. Once packed and covered with liquid, the fruit is processed *(pages 30-31)* in the same way as tomatoes *(pages 32-33)*.

1 Halving the fruit. Wash fruit—here, greengage plums. With a paring knife, cut through the flesh of the fruit down to the pit and all the way around it. Twist the two halves of the fruit in opposite directions to separate them. Lift out the pit with your fingers or, if necessary, pry it out with the tip of the knife.

2 Removing fibers. With the tip of the knife, cut and scrape the cavities of the plum halves to remove all of the tough, fibrous flesh that surrounds the pit. Place the halved, pitted plums in a bowl.

4 Packing the jars. Wash canning jars and turn them upside down on a towel. Fill a jar halfway with fruit. Add one or two kernels and continue packing the jar with plums to within ½ inch [1 cm.] of the top—a quart [1-liter] jar will hold six or seven plums.

3 **Extracting the kernel.** When all of the plums have been halved and scraped, crack open one or two pits with a nutcracker *(above, left)*. Use the tip of the knife carefully to pry the kernel from the pit *(center)*. Peel away the inner skin that covers each kernel *(right)*. If the skins do not come free easily, loosen them by blanching the kernels in boiling water for one minute.

5 **Covering with syrup.** Prepare a medium syrup. Boil the syrup for one minute to drive out any air bubbles, which could discolor the preserved fruit. Remove the syrup from the heat and ladle it through a canning funnel to cover the plums in each jar, leaving ½ inch [1 cm.] of headspace. Cover, and process the jars in a boiling water bath.

Insurance against Bursting

Piercing the skin. To prevent the skins of whole plums *(above)* or other tough-skinned fruits from bursting in the heat of the processing bath, prick each fruit once lightly with a skewer or other pointed implement. After pricking, pack and process the fruit just as you would sectioned fruits.

Advance Cooking for Tenderness and Taste

Hot packing—a term that simply means cooking food before you pack it into canning jars and process it *(pages 30-31)*— serves two purposes. The cooking softens fibrous or very firm foods, thereby making them more palatable. And cooking also allows you to imbue food with extra taste and color; you need only add herbs, spices or other flavorings to the pot.

Hot-packing preliminaries depend on the nature of the food being preserved: The cooking can range from brief parboiling to long simmering *(guide, pages 30-31)*. In the top demonstration, for example, rhubarb is first sugared, then left to steep and render its juice, which is used to cook the rhubarb briefly—just long enough to tenderize it *(recipe, page 148)*. Firmer foods, such as pears *(right, bottom; recipe, page 151)*, require gentle and relatively lengthy cooking in liquid to soften their flesh.

Like the rhubarb—or any other fruit —the pears should be cooked with sugar, which helps them retain their texture. The shape of the pears dictates that the sugar be in the form of a syrup to ensure that all of their smooth surfaces are covered completely; in this case, the pears' sugar syrup is made with red wine, which gives the fruit a particularly appealing flavor and color.

Spices and herbs lend distinction to either type of preparation. When the cooking time is short, the spices should be ground so that they will contribute their flavors quickly. Spices for longer-cooking foods can be used whole: To prevent the whole spices from dispersing through the cooking liquid, you can enclose them in a cloth bag before adding them to the pot *(box, below)*; the bag can be removed after the spices it contains have rendered their flavors.

Poaching in Natural Juice

1 **Cutting the rhubarb.** Cut away and discard the poisonous leaves from stalks of rhubarb. Rinse the stalks, pat them dry, and cut them into pieces 1 to 2 inches [2½ to 5 cm.] long. Place the pieces in a mixing bowl and toss them with sugar.

Fashioning a Spice Bag

Whole spices release their flavorful oils more slowly than ground ones. But when given enough time to act—either by long cooking or by marination in a pickling or brining solution— whole spices work just as well and, unlike ground spices, they will not cloud the liquid in which they are placed.

Whole spices can be dropped directly into the preserving liquid. To keep the liquid free of bits of spice, tie the spices in a muslin or cheesecloth bag and remove after cooking.

The spices you use depend entirely on the food you are preserving: Whole cloves, cinnamon sticks, allspice and sliced fresh ginger are combined for use in the pear-and-wine combination shown at right. Before being wrapped, large pieces of hard spices such as allspice should be bruised with a mortar and pestle to release some of their oils. Softer spices such as ginger should be partially crushed with a blow from the flat of a knife blade.

Filling a spice bag. Cut a 6- to 8-inch [15- to 20-cm.] square of cheesecloth or muslin, shown here. Place bruised or partially crushed spices in the center. Gather the corners to form a bag, loop a string around the neck and tie it. Knot the trailing ends to form a loop to attach to a pot handle.

Simmering in a Wine Syrup

1 **Peeling the pears.** Pour wine into a nonreactive pot. With a vegetable peeler, remove skins from hard, green pears—leaving the stem on—and immediately place the pears in the wine to keep them from turning brown. Loop the string of a spice bag *(box, left)* around the pot handle and drop the bag into the wine.

2 **Grinding the spices.** Let the rhubarb steep in the sugar at room temperature for four hours. Grind whole allspice and whole cloves with a mortar and pestle, and set them aside in a small dish. Use a nutmeg grater to grate a nutmeg, then blend it with the other spices.

3 **Precooking the fruit.** Transfer the rhubarb and the syrup formed by the rendered juice to a nonreactive pot. Add the spices and stir the mixture with a wooden spoon. Boil the mixture over high heat for one or two minutes, until the rhubarb can be pierced with a knife. Remove the pan from the heat.

4 **Packing and processing.** Ladle the rhubarb mixture into clean, hot jars through a canning funnel, leaving a ½-inch [1-cm.] headspace. Add a 1-inch [2½-cm.] piece of stick cinnamon to each jar. Close the jars and process in a boiling water bath.

2 **Poaching the pears.** Place the pot over high heat. Add sugar and bring the mixture to a boil. Reduce the heat and simmer, uncovered, for 20 minutes, until the pears are tender enough to be pierced with a knife. Remove the pot from the heat and let the pears and wine cool for 30 minutes. Remove the spice bag.

3 **Packing the jars.** Prepare canning jars for packing (pages 10-11). Using a spoon or your fingers, set a pear in a jar with its stem up. Pack the next pear with its stem down. Continue packing, alternating stem ends to ensure a snug fit, and leave a 1-inch [2½-cm.] headspace.

4 **Topping with wine.** When all of the pears have been packed, return the wine syrup to the heat and boil it rapidly, uncovered, until it has reduced by half. Using a funnel and ladle, fill each packed jar with syrup, leaving a ½-inch [1-cm.] headspace. Process (pages 30-31) in a boiling water bath.

Juices for Drinks and Syrups

Juices—the bases for delicious sauces and syrups or ready-made drinks—are made by extracting the liquid from fruits and vegetables, then straining and flavoring it. The quickest way to obtain juice is with a manual or electric juice extractor; several different types are sold at kitchen-equipment shops. But, as seen in the two demonstrations here, no special tools are needed to extract the juice.

To remove liquid from very soft fruit, you can simply crush the fruit, let it sit for a day or so to render its juice, then strain the juice to remove seeds and pulp. The technique—applied to raspberries in the top demonstration *(recipe, page 93)*—is a good way to preserve the essences of very ripe fruits that are too fragile to be canned whole. Leftover pulp may then be used to make fruit butters and cheeses *(pages 58-59)*.

Firm or fibrous fruits and most vegetables require more radical treatment than crushing: They are chopped and cooked to break down their tissues and release moisture. In the bottom demonstration *(recipe, page 92)*, green peppers, celery and onions are simmered together with tomatoes and seasonings; then the liquid is strained and pressed from the pulp.

Once fruit or vegetable juice has been extracted, it may be flavored in a number of ways. The raspberry juice shown here receives a brief final cooking with sugar to make a light syrup that can be used as a sauce for ice cream or to make drinks. The juice could also have been enhanced with flavorings such as almond extract or lemon juice. The vegetable juice in the bottom demonstration was mixed with lime juice, which contributes a sharp, piquant flavor.

Fruit juices, sealed in canning bottles or jars, are acidic enough to process in a boiling water bath *(pages 32-33)*. Vegetable juices made primarily of tomatoes may also be processed this way, but recipes for these juices must be followed strictly: Inclusion of too great a quantity of other vegetables can lower the tomatoes' acidity to the point at which processing the juice in boiling water will not preserve it safely. In such a case, the juice must be processed in a pressure canner *(pages 40-41)*.

A Manual Method for Soft Fruit

1 **Crushing the berries.** Pick over fresh raspberries and remove any that are underripe or spoiled. Rinse the berries. Place the berries in a large bowl and crush them lightly with a pestle. Cover the bowl and refrigerate for a day.

2 **Straining the raspberries.** Drape a doubled piece of cheesecloth or, as here, a large kitchen towel over a second bowl; press the cloth down to line the bowl. Tip the raspberries and their juice into the cloth-lined bowl.

A Preliminary Cooking for Firm Produce

1 **Chopping vegetables.** Coarsely chop tomatoes, onion, green pepper and celery, and place them in a nonreactive pot. Stir in mustard seeds, black peppercorns, a dried chili and coriander. Wash a few sprigs of basil and parsley, peel a garlic clove, and add them to the mixture.

2 **Softening the vegetables.** Bring the mixture to a boil, stirring occasionally with a wooden spoon to prevent sticking. Reduce the heat and simmer, uncovered, until the vegetables are soft and juicy—about 30 minutes.

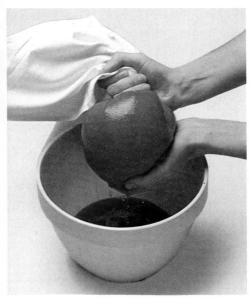

3 **Extracting the juice.** Gather up the four corners and twist them together so that the cloth encloses the fruit in a pouch. Holding the ends of the cloth in one hand, twist with the other to squeeze out the juice. If you like, you may bottle and process the juice at this point.

4 **Making a syrup.** Transfer the juice to a nonreactive pot. Stir in sugar. Heat the liquid gently, stirring to dissolve the sugar. Bring the mixture to a boil, then take the pan off the heat and remove any scum from the surface.

5 **Canning.** Ladle the raspberry syrup through a funnel into canning bottles, leaving a ½-inch [1-cm.] headspace. Cover the bottles and process in a boiling water bath.

3 **Straining.** Place a food mill, fitted with a fine disk, over a large bowl. Ladle the tomato mixture into the mill, a few cups at a time. Turn the handle of the mill until juice stops flowing through the disk. Discard the pulp, skins and seeds.

4 **Skimming the liquid.** Cover the bowl of juice and refrigerate it for at least six hours to let it settle. With a ladle or large spoon, skim off the watery liquid that has collected atop the juice. Save the liquid for soup stock. Pour the remaining juice into a nonreactive pot.

5 **Flavoring the juice.** Place the pot containing the juice over medium heat. Squeeze lemons or limes and strain their juice. When the tomato juice is warm, add salt, sugar, and lemon or lime juice to taste. Pour the mixture into canning bottles and process in a boiling water bath.

High Temperatures for Low-Acid Foods

Meats and all vegetables other than tomatoes lack the acid that neutralizes the deadly botulism spores present in foods. However, low-acid foods can be safely canned by processing them with a heat high enough to kill the spores—at least 240° F. [116° C.].

A boiling water bath will not suffice: Under normal atmospheric pressure, water boils at a temperature level of 212° F. [100° C.], and it boils away into steam. However, if the pressure on the water is raised by containing the steam, its boiling point will be increased accordingly. An additional 10 pounds of pressure per square inch [70 kPa] allows boiling water to reach 240° F.

In preserving, this elevation of the water's boiling point is achieved by means of a pressure canner. The device consists of a heavy deep pot, fitted with a rack to hold jars, and a lid that can be secured to prevent the escape of steam. To prevent the pressure from rising too high and blowing off the lid of the canner, the lid of the pot is equipped with a pressure or weight gauge and a means to vent steam; in addition, it has a safety valve or plug that releases steam in case the internal pressure is allowed to rise too high.

When using a pressure canner, take special care to follow the manufacturer's instructions for letting steam and air escape in order to maintain the necessary pressure of 10 pounds. Be sure to adjust processing times for altitude (page 31).

Both the raw-packing and the hot-packing methods explained on pages 32-37 can be used to load jars for pressure canning. Most vegetables are customarily packed raw to preserve as much texture, color and flavor as possible; in the top demonstration, however, beets are scalded to loosen their tough skins. Meat, poultry and game are tastier if precooked and hot-packed. In the bottom demonstration, beef is braised in beef stock, which then serves as the packing liquid.

Cooking Beets in Two Stages

1 **Trimming the beets.** Trim off beet greens, leaving 2 inches [5 cm.] of the stems above the beets and the taproots below. Take care not to cut into the beets themselves, which would release their juice, color and flavor. Scrub the beets under running water and drop them into a large pot of boiling water; boil the beets for 20 minutes.

2 **Skinning the beets.** Transfer the beets to a colander and run cold water over them to stop the cooking. Set the colander over a bowl to drain. With your fingers, pull off the stem left on each beet, squeezing the beet as you do so to force it out of its skin. Cut out any black or hard spots.

Canning Beef with an Enrichment of Stock

1 **Browning beef cubes.** Trim fat from meat — a beef round roast, in this case — and cut the meat into cubes. Cooking small batches at a time, brown the cubes in a little oil over medium heat, transferring each batch to a bowl as it is finished (above, left). When all of the cubes have been browned, return them to the pot and cover with beef stock. Partly cover the pot and simmer the meat for 15 minutes, stirring occasionally (right).

3 **Packing jars.** Pack the peeled beets into prepared canning jars *(pages 10-11)* to within 1 inch [2½ cm.] of the tops; it may be necessary to cut some in half. Pour boiling water into the jars, leaving a ½-inch [1-cm.] headspace. Add ½ teaspoon [2 ml.] of salt to each jar, if desired for flavor, and close the jars.

4 **Filling the canner.** Pour water into a pressure canner to a depth of 2 inches [5 cm.] and lower the jars onto the rack in the canner. Secure the lid. Set the canner over medium heat and bring pressure to 10 pounds per square inch [70 kPa]; here, a weight gauge that controls escaping steam is lowered onto a canner's vent *(inset)*. When the pressure is correct, begin counting processing time. When the time is up, cool the canner before removing the jars.

2 **Skimming the fat.** Using tongs, transfer the cooked meat cubes to prepared canning jars, packing them loosely to within 1 inch [2½ cm.] of the jar rims. Skim surface fat from the stock with a spoon, using a double thickness of paper towel to blot up any fat the spoon does not remove.

3 **Filling the jars.** With the aid of a canning funnel, ladle the stock into the jars, leaving about 1 inch [2½ cm.] of headspace. Cover the jars and process them in the pressure canner, following the manufacturer's directions.

3
Jellies and Jams
Alliances of Fruit and Sugar

An extraordinary range of preserves—jellies, marmalades, jams, conserves, fruit butters and fruit cheeses—are made by cooking fruit with sugar. Sugar may seem an unlikely preserving agent: Microorganisms thrive on it in weak solutions. But in strong concentration, sugar has a dehydrating effect, similar to that of salt, which inhibits the development of microorganisms so effectively that properly sealed fruit preserves can last for a year or more. To ensure long storage life, most such preserves are made with ¾ to 1 cup [175 to 250 ml.] of sugar for each cup of fruit juice or prepared fruit.

In translucent jellies and clear citrus marmalades made from fruit juice, sugar not only keeps the food from spoiling but also helps it set. The fundamental jelling agent is pectin—a gumlike substance yielded by the flesh, skins and seeds of most fruits when they are boiled. But only if the fruit has adequate pectin, and only if both sugar and acid are present, will the jellies or marmalades set to the proper firmness. It is possible to make pectin stock *(page 45)* or to buy commercial pectin to supplement that of low-pectin fruits, and acid can be added to low-acid fruits in the form of lemon juice. The chart on page 44 shows which fruits contain enough natural acid and pectin to jell unaided, and which need additional quantities of one or the other.

The jelling power of pectin is less important in jams and conserves, made from crushed fruit, and fruit butters and cheeses, made from puréed fruit. These preserves should be thick and soft, rather than firm. Some fruit preserves do not need to jell at all. In one type of jam, fruit is simply boiled in a heavy sugar syrup for as brief a time as is necessary to cook it. Once the fruit is done, the syrup can be reduced to provide the concentration of sugar needed for preservation. The strawberry preserve shown opposite is one example: Minimum cooking gives it a fresher flavor than that of more orthodox jams.

Once made and poured into sterile jars, jellies need only be covered with paraffin *(page 11)*—or jar lids—and stored in a dry, cool, dark place. However, fruit butters, fruit cheeses, and sweet preserves that contain solid pieces of fruit all will spoil quickly. To kill the bacteria that can cause molds, such preserves should be processed in a water bath for the time periods specified in the guide on pages 30-31.

A batch of strawberries, boiled briefly in sugar syrup, is transferred to a colander to drain. The rich, red syrup will be boiled to reduce its volume before the next batch of berries is added; finally, all of the fruit will be returned to the pan so it can absorb the syrup *(page 54).*

The Factors in Setting

The consistency of all fruit jellies and that of some jams is determined by the balance among three ingredients: sugar, acid and pectin.

The amount of sugar remains fairly constant for virtually all fruit preserves: For effective preservation, these mixtures must contain 1½ to 2 cups [375 to 500 ml.] of sugar for every 2 cups of prepared fruit or fruit juice. The levels of acid and pectin, however, vary greatly from fruit to fruit, as indicated in the chart below. Even the same kind of fruit will have different acid and pectin levels according to variety and to age: Under-ripe fruit is highest in acid and pectin. Preserves made from fruit that is rich in both acid and pectin—tart apples, oranges and cranberries, for example—set easily. But if a fruit is low in either

substance, a preserve made from it must include supplementary acid or pectin.

The acidity of a specific batch of fruit is readily judged by taste: A sharp or tart flavor is a good sign of acidity. The presence of pectin is not so apparent, but when the fruit is cooked and made into juice, its pectin level can be determined by a simple test (box, opposite).

Acidity is easily boosted by adding lemon juice to the fruit before cooking. The juice of one lemon—about 2 tablespoons [30 ml.]—is sufficient for 2 cups of low-acid prepared fruit or fruit juice.

There are three ways to supplement pectin. You can include in the preserve mixture another fruit rich in pectin. If you want a jelly tasting of only one fruit, you can use commercially prepared pectin, but this requires adding so much sugar to the mixture that it yields a jelly too sweet for many palates.

The best way to get a fine, distinctive-tasting jelly from low-pectin fruit is to reinforce the juice with homemade pectin stock, which is produced by boiling high-pectin fruit until it renders its juice, straining off the juice, and reducing its volume until the pectin is concentrated enough to set the liquid (right). As a rough guide, ⅔ cup [150 ml.] of the stock will be enough to set 4 cups [1 liter] of low-pectin prepared fruit or fruit juice.

The pectin stock demonstrated here is made with apples, whose unassertive flavor will not mask other fruits' qualities. Quinces are equally adaptable. However, stocks made from assertively flavored high-pectin fruits, such as gooseberries, cranberries or lemons, should be used only when they are to play a part in the taste of the finished preserve.

An Analysis of Jelling Abilities

	PECTIN			ACID		
	HIGH	MEDIUM	LOW	HIGH	MEDIUM	LOW
Apples: Sweet		•				•
Apples: Tart	•			•		
Apricots			•	•		
Blackberries		•		•		
Blueberries		•		•		
Cherries: Sour		•			•	
Cherries: Sweet		•				•
Citrus Fruits	•			•		
Cranberries	•			•		
Currants	•			•		
Elderberries		•			•	
Figs			•			•
Gooseberries	•			•		

	PECTIN			ACID		
	HIGH	MEDIUM	LOW	HIGH	MEDIUM	LOW
Grapes		•		•		
Guavas		•		•		
Melons		•				•
Nectarines		•				•
Peaches		•				•
Pears		•				•
Pineapple		•	•			
Plums: Sour	•			•		
Plums: Sweet		•			•	
Quinces		•				•
Raspberries		•		•		
Rhubarb			•	•		
Strawberries			•		•	

How to use the chart. Locate the fruit you want to use for jelly and read across the chart. Dots in the appropriate columns indicate the fruit's approximate level of acid and pectin. If a fruit is low in either acid or pectin, these components must be supplemented to enable the juice to set after it is cooked with sugar.

A Homemade Pectin Stock

1 **Cooking the fruit.** Put chopped apples in a nonreactive pan. Add a little water, cover and simmer until soft — approximately 20 minutes.

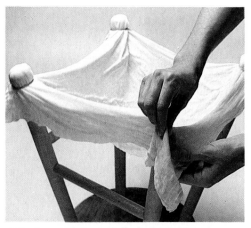

2 **Improvising a jelly bag.** Fold a piece of heavy cheesecloth into six layers and tie the corners to the legs of an upturned stool *(above)*.

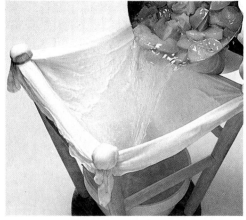

3 **Straining the juice.** Place a bowl under the bag. Pour the apples and their liquid into the bag. Without stirring, let the juice drain overnight.

4 **Reducing the liquid.** Pour the strained juice into a pan. Bring the juice to a boil and cook until it has reduced to about half its original volume.

5 **Straining for clarity.** Test for pectin *(box, below)*; if necessary, continue reducing and testing. Strain the juice through a double layer of cheesecloth.

6 **Canning the stock.** Fill warm, clean canning jars with the stock. Cover the jars and process *(pages 30-31)* in a boiling water bath.

Testing the Pectin Level

During boiling, water evaporates from pectin stock, concentrating the pectin to the point where the stock can serve as a setting agent for jellies made from low-pectin fruit *(page 44)*. To determine when this point has been reached, mix a small sample of stock with rubbing alcohol, available at drugstores. The alcohol interacts with the stock, causing it to clot if the pectin level is high enough. Do not taste this sample, and do not return it to the stockpot: Rubbing alcohol is poisonous.

1 **Mixing stock and alcohol.** Pour 1 teaspoon [5 ml.] of pectin stock onto a plate; cool. Add 2 tablespoons [30 ml.] of rubbing alcohol. Swirl the plate until clots begin to form.

2 **Judging pectin.** Juice high in pectin forms a coherent mass *(above)*; a medium pectin level is indicated by several large clots, weak pectin by small, scattered clumps *(page 48, Step 5)*.

A Self-setting Jelly Enhanced with Herbs

Jellies are made in much the same way as pectin stock *(page 45):* Fruit is cooked in water until it renders its juice, then the juice is strained off and boiled down. The addition of sugar during the boiling makes the juice set. Sugar interacts with the pectin and acid in the juice so that the mixture jells when it cools after cooking.

This procedure is simplicity itself, although care must be taken to ensure that the liquid not only sets to the proper quivering firmness, but also that it remains unclouded, and fresh in flavor. Although any fruit juice can be made to jell by the addition of pectin, the easiest course is to use fruits that themselves contain sufficient pectin and acid *(chart, page 44).* Among them are apples, cranberries, currants, and the lemons and oranges used here. Such fruits can be used alone or in combination, and all may be flavored with fresh herbs such as sage, mint or—as shown—rosemary.

For the richest flavor, boil the fruit with a minimum of water—just enough to cover the bottom of the pan and prevent sticking. To produce a clear, sparkling jelly, let the juice drip unaided from the cooked fruit *(Step 3):* Crushing the fruit would cloud the jelly.

When making juice, you can cook any amount of fruit you wish, but boil only small quantities of the juice at a time to make jelly—6 cups [1½ liters] at the most. Larger quantities take so long to reduce to jelling consistency that the juice might overcook and its sugar crystals cluster in grains, clouding the finished jelly and ruining its texture.

As water evaporates from the boiling juice, the liquid's temperature will rise, indicating that the pectin is becoming concentrated enough to jell. The most accurate way to determine when the jelling point has been reached is to measure the temperature of the juice with a candy or jelly thermometer: When the temperature reaches 8° F. [5° C.] above the boiling point at your altitude—220° F. [105° C.] at sea level—the juice has become a liquid jelly. To be sure, observe the jelly when it is dropped from a spoon *(Step 5).* For extra safety, test the jelly by chilling a spoonful to see if it sets *(box, opposite).* Stop cooking as soon as the jelling point is reached: Overcooking toughens jelly.

1 **Preparing the fruit.** Cut oranges in half and cut the halves into slices ⅛ inch [3 mm.] thick. Put the slices in a large bowl. Squeeze the juice from lemons; measure the juice and add it to the bowl. Add three times as much water as you have juice and stir. Cover the bowl and refrigerate it overnight or for at least eight hours.

2 **Cooking the fruit.** Pour the fruit and juice into a large, nonreactive pot, set the pot over medium heat and bring the mixture to a boil. Boil it, uncovered, for 25 minutes. Then stir in fresh rosemary leaves and let the fruit boil for five minutes more.

5 **Testing for jell.** When the juice's temperature reaches 220° F. [105° C.], use a clean, chilled spoon to lift out a little liquid. Hold the spoon at least 12 inches [30 cm.] above the pot to avoid the steam, and let the liquid fall from the spoon. If it falls in two drops that move together to form a sheet *(inset),* the jelly is ready.

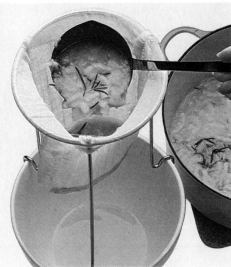

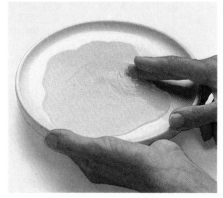

Chilling: An Optional Test

Extra insurance. While the jelly cooks, chill a clean plate in the freezer. After about 15 minutes, when the jelly temperature has risen to 220° F. [105° C.], spoon a little jelly onto the plate and return the plate to the freezer for one or two minutes. Then push the jelly with your fingers. If the jelly wrinkles as it is pushed, the jelling point has been reached.

3 **Straining out juice.** Remove the pot from the heat, cover it and let the mixture steep for half an hour. Suspend a jelly bag in its frame over a large bowl. Pour the fruit and juice into the bag and drain for about four hours; do not press on the fruit. Measure the juice, pour it into a nonreactive pot and boil for two or three minutes.

4 **Cooking the jelly.** Stirring constantly, add sugar to the boiling juice, allowing 1¾ cups [425 ml.] of sugar for each 2 cups [½ liter] of juice. Insert a jelly thermometer into the pot and continue to boil the juice for about 15 minutes, stirring occasionally to prevent it from boiling over.

6 **Removing scum.** Remove the jelly from the heat and, with a long-handled spoon, skim off any surface scum, which could cloud the jelly. Use a ladle to immediately pour the hot jelly into hot, sterilized canning jars, leaving ½ inch [1 cm.] of headspace. Seal the jars with paraffin *(page 11)*. Let the paraffin cool and harden.

7 **Decorating the jelly.** Cut fresh rosemary into small sprigs. Hold each sprig with tweezers and dip the base into melted paraffin. Let the paraffin cool a moment, then press the sprig base onto the cool paraffin that covers the jelly. With a spoon, place a drop of hot paraffin on the base of the sprig to secure it to the paraffin top.

Making Jelly from Low-Pectin Fruit

Delicious jellies can be made with low-pectin fruits such as pears or strawberries *(chart, page 44)* if the pectin content in the fruit juice is raised to the point where it will jell. One tactic is to make a mixed jelly that includes high-pectin fruit; you could combine pears with fresh currants, for instance, or strawberries with limes. The only requirement is that the high-pectin fruit account for at least three quarters of the juice volume.

If you prefer the purer taste of a single-fruit jelly, you can add any necessary pectin at a later stage of cooking in the form of pectin stock. Often, the need for this addition will be in doubt until a sample of the juice is tested with rubbing alcohol *(page 45)*. Such is the case with grapes, a traditional favorite of jelly makers *(recipe, page 96)*; their pectin content varies considerably, depending on the grape variety and on ripeness. If, as in this demonstration, the juice fails the test *(Step 6)*, add the pectin stock before mixing in the sugar and completing the cooking.

1 **Crushing the fruit.** Wash grapes under cold running water and strip them from their stems. A small batch at a time, place the grapes in a bowl and crush them with a pestle to rupture the skins. Transfer each crushed batch to a large, nonreactive pot — in this case, enameled cast iron.

2 **Simmering the grapes.** Set the pot of grapes over medium heat. Stir a little water into the fruit to prevent sticking. Stirring occasionally, bring the mixture slowly to a boil.

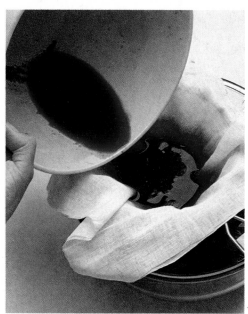

5 **Removing sediment.** Line a strainer with a double layer of cheesecloth. Set the strainer over a large, nonreactive bowl. Taking care not to disturb the sediment, pour the juice through the strainer. Discard the sediment.

6 **Testing for pectin.** Measure the juice into a pot. Spoon 1 teaspoon [5 ml.] of juice onto a plate and add 2 tablespoons [30 ml.] of rubbing alcohol. Swirl the liquids. If the pectin is adequate, the juice will immediately clot into a large clump. If only small, separate grains appear *(inset)*, add ⅔ cup [150 ml.] of pectin stock for every 4 cups [1 liter] of juice.

3 **Finishing cooking.** Boil the grapes, uncovered, for 15 to 20 minutes, or until they have rendered most of their juices. Stir from time to time to prevent the juices from boiling over. Then remove the pot from the heat and allow the fruit mixture to cool slightly.

4 **Straining out juice.** Suspend a jelly bag in its frame over a large, nonreactive bowl. Fill the bag with the cooked grapes and their juice, and let the juice drip through undisturbed for three to four hours. Cover the bowl, and chill the juice for at least eight hours so that the sediment caused by the grapes' tartaric acid can settle; the sediment would otherwise make the jelly granular.

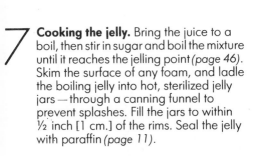

7 **Cooking the jelly.** Bring the juice to a boil, then stir in sugar and boil the mixture until it reaches the jelling point (page 46). Skim the surface of any foam, and ladle the boiling jelly into hot, sterilized jelly jars — through a canning funnel to prevent splashes. Fill the jars to within ½ inch [1 cm.] of the rims. Seal the jelly with paraffin (page 11).

Marmalades: Suspensions of Peel and Pulp

Marmalade is fruit jelly that incorporates the fruit's skin and flesh. Most marmalades are based on oranges, which are high in pectin and impart a distinctive bittersweet taste to the preserve; the oranges may be used either alone, in combination with other citrus fruits, or with such noncitrus fruits as pineapple. The mixture can be flavored with liqueurs, whisky or spices. And, as indicated by the two demonstrations at right, the texture of the marmalade can be varied at will, simply by altering the preliminary preparations used to soften the fruit's tough skin before sugar is added and the mixture is boiled to the jelling point.

To produce a marmalade full of firm, chunky fruit pieces *(top demonstration; recipe, page 100)*, the fruits are cooked whole in a little water for about an hour. Then the fruits are chopped and—with the pectin-rich seeds, tied in a muslin bag—boiled in a sugar syrup made from the cooking liquid. The relatively brief cooking process leaves the fruit with a pleasant, chewy texture, a fine contrast to the smooth jelly that surrounds it.

A more time-consuming technique is used to produce a delicate, clear marmalade with small shreds of tender peel *(bottom demonstration; recipe, page 101)*. The fruits are sliced thin, soaked overnight, then cooked—with the seeds—in the soaking water for about two hours. Only then is sugar added and the cooking completed. The soaking and lengthy cooking give the peel an almost melting consistency, and the large amount of water attenuates the fruits' tartness.

A Chunky Blend with a Bitter Edge

1 **Cooking the fruits.** Wash the fruits — here, oranges and a lemon. Put the fruits in a large, heavy, nonreactive pan. Add water, cover, and boil for an hour, until the fruit is easily pierced with a skewer.

2 **Straining the fruits.** Place a colander over a large bowl. Transfer the fruits and their liquid from the pan to the colander, letting the liquid drain into the bowl beneath. Reserve the liquid.

A Mild Preserve with Tender Peel

1 **Soaking the fruits.** Wash fruits — oranges, lemons, grapefruit and limes are shown — and slice the fruits thin. Remove the seeds and tie them in a muslin or cheesecloth bag. Put the fruits and seed bag in a large bowl, and add water. Cover the bowl with plastic wrap and allow the fruits to soak overnight in a cool place.

2 **Cooking the fruits.** Tip the fruits, seeds and water into a large, heavy, nonreactive pan. Bring the fruits to a boil over medium heat, then simmer for about two hours, uncovered, until the pulp is clear and the peel is tender enough to slice with the edge of a spoon.

3 **Chopping the fruits.** Cut the fruits into quarters. Use the tip of the knife to remove the seeds; set them aside. Chop the fruits into equal-sized pieces — in this case, they are cut coarse to produce a chunky marmalade. Wrap the seeds in a muslin or cheesecloth bag *(page 36)*.

4 **Cooking with sugar.** Pour the reserved cooking liquid back into the pan and add sugar. Stir the mixture over low heat to dissolve the sugar, then boil the syrup for about five minutes. Use a metal spoon to remove any scum. Stir in the chopped fruits and add the bag of seeds.

5 **Filling jars.** Boil and stir until the jelling point is reached *(page 46)*, about 30 minutes. Then discard the seed bag and skim the marmalade. Allow it to cool a moment, then stir to distribute the peel. Using a ladle and a funnel, fill warm, clean canning jars to within ½ inch [1 cm.] of the rims. Cover, and process *(pages 30-31)* in a boiling water bath.

3 **Finishing the marmalade.** Add sugar. Stir the mixture until the sugar is dissolved. Increase the heat and boil the marmalade rapidly, stirring occasionally, for about 30 minutes, until the jelling point is reached *(page 46)*. Remove the pan from the heat and lift out the seed bag. Skim the marmalade. Allow it to cool slightly.

4 **Canning the marmalade.** Stir the marmalade to distribute the peel. Using a funnel, ladle the marmalade into warm, clean canning jars, leaving ½ inch [1 cm.] of headspace. Cover, and process in a boiling water bath.

Jams: Simple Blends of Crushed Fruit and Sugar

Of all fruit preserves, none is simpler to make than jam: Fruit is crushed, then boiled with sugar in its own juices until it has reduced to a thick, soft mass. Although fruit rich in pectin will thicken faster, the amount of pectin in the fruit is not critical. Jams are not juices that must jell, but solid pieces of fruit suspended in liquid; as cooking progresses, the liquid evaporates, thickening the mixture.

Virtually any fleshy fruit—apricots, berries, plums or peaches, for instance—lends itself to jam making. The fruit should be ripe but still firm. Soft, overripe fruit would disintegrate during the boiling and rob the jam of its varied consistency. Most fruits other than berries should be peeled and pitted before cooking. Once peeled, pale-colored fruits should be dropped into acidulated water *(pages 14-15)* to prevent their flesh from darkening before cooking begins.

Besides the sugar necessary for preservation, jams may be cooked with any flavoring you desire—brandies, liqueurs or, as in the peach jam demonstrated here, spices *(recipe, page 106)*.

1 **Preparing peaches.** Drop peaches into boiling water for one minute to loosen their skins, then plunge them into cold water to stop the cooking. Peel and halve the peaches, remove their pits, and scrape out any dark fibers from the hollows. Drop the halves into water that has been acidulated with a little fresh lemon juice. A small batch at a time, place the peach halves in a bowl and crush them with a potato masher *(above)* or a pestle.

4 **Testing for doneness.** When the jam is ready it will be thick in consistency and dark in color; the fruit pieces will be translucent. As a test of consistency, lift a spoonful: The jam should be thick enough to mound slightly on the spoon. Remove the pan from the heat. Do not overcook the jam: It may thicken further as it cools.

5 **Extracting spicy flavors.** With the spoon, lift the spice bag from the jam and place it in a small heatproof bowl. Tilt the bowl over the pot of jam and press the bag with the back of the spoon to force its flavored juices to drip into the jam. Discard the spice bag.

2 **Cooking the fruit.** As each batch of peaches is crushed, transfer it to a nonreactive pot. When all of the fruit is crushed, set the pot over medium heat and stir in a little water to prevent sticking. Stirring from time to time, bring the mixture to a simmer and cook, uncovered, for 10 minutes, or until the fruit is tender.

3 **Boiling the jam.** Prepare a spice bag containing whole cloves, whole allspice and a broken cinnamon stick. Stir sugar into the simmering fruit and, when the sugar dissolves, add the spice bag. Increase the heat and bring the mixture to a boil. Boil rapidly for about 15 minutes, stirring frequently — constantly toward the end of cooking — in order to prevent scorching.

6 **Canning the jam.** Use the spoon to skim any foam from the surface of the jam into a small heatproof bowl. Discard the foam. Ladle the jam through a canning funnel into hot, clean canning jars, leaving a ½-inch [1-cm.] headspace. Process the jars (pages 30-31) in a boiling water bath.

A Cooking Strategy with a Reviving Effect

Jams require lengthy cooking to evaporate fruit juices and reduce them to the proper thick consistency. Yet the longer a fruit is cooked, the less natural color, taste and texture it will retain. To ensure that the fruit remains as close to its natural state as possible, many cooks prefer to simply boil it in sugar syrup until it becomes tender, then can it as a preserve, which may be used in the same way as jams and jellies.

An elaboration of this method is shown here with preserved strawberries (recipe, page 110); the procedure is uniquely suited to these and other thin-skinned berries—raspberries, for instance—and makes maximum use of their juice and their sweetness.

First, a small batch of strawberries is boiled in a sugar syrup for about one minute. This initial cooking draws the juice—and color—into the syrup; the berries, limp and grayish at this stage, are set aside to drain. The syrup is reduced by boiling to concentrate its color and flavor. Then another batch of berries is added to the syrup and the procedure repeated. When all of the strawberries have been boiled in this way, the juices that have drained from them are added to the syrup, which is then reduced to its original volume.

Finally, all the strawberries are boiled again with the reduced syrup. During this second cooking, the berries undergo a remarkable transformation: They absorb much of the deeply colored liquid and become red and plump once again, ready to be transferred from the pan to canning jars.

The key to the process is the repeated reduction of the juice-enriched syrup. As a result of the reduction, the fruit will yield only about half its weight in preserves, but the intensity of flavor is extraordinary. The reduction also concentrates the natural sugar drawn from the strawberries; the only additional sugar needed is the small amount in the syrup for the first cooking.

The preserves will keep for six months in the refrigerator; if you wish to store them longer, process them (pages 30-31) in a boiling water bath.

1 **Washing strawberries.** Pick over small ripe berries and discard any that are damaged or underripe. Put them in a bowl of fresh water. Swirl the berries rapidly in the water with your hands, then lift them out with splayed fingers and drain them in a colander set over a tray.

2 **Removing the hulls.** To prevent loss of juice and flavor, remove the hulls only after the berries are washed. Pull the hulls and cores from the berries. If the core does not come free with the hull, leave it in the fruit: It will not affect the quality of the preserves. Put the fruit on a tray.

4 **Adding berry juice.** Repeat the cooking, draining and reducing process until all of the fruit has been cooked and transferred to the colander. Holding the colander over the pan to catch any dripping berry juice, tip the drained juice in the pie plate into the syrup.

5 **Cooking the berries.** Boil the juice-enriched syrup for about five minutes more to reduce it to its original volume. Then tip the drained berries back into the syrup. Increase the heat to bring the syrup back to a simmer, and simmer for five minutes, stirring very gently with a wooden spoon to avoid crushing the berries.

3 **Cooking the fruit.** In a large, nonreactive pot, make a sugar syrup; for each 3 pounds [1 ½ kg.] of fruit allow 1 cup [250 ml.] each of sugar and water. Stir the sugar and water over low heat to dissolve the sugar, then bring the syrup to a boil. Drop about 1 pound [½ kg.] of berries into the syrup *(above, left)*. After the syrup returns to a boil, cook the fruit for one minute. Using a perforated spoon or skimmer, transfer the berries to a colander set over a pie plate to catch drips *(center)*. Stirring occasionally, boil the syrup for about five minutes to reduce it to its original volume *(right)*.

6 **Finishing the preserves.** When the berries have absorbed most of the syrup and are thickly coated with the rest *(inset)*, remove the pan from the heat. Ladle the strawberry preserves into hot, clean canning jars *(right)*. Cover the jars and process in a boiling water bath.

Conserves: Imaginative Combinations of Fruit and Flavorings

By marrying the flavors of fresh fruits with those of nuts, dried fruits, spices, liqueurs and other spirits, you can produce a range of unusual jams. Such conserves—as these mixtures are frequently called—have several roles: They can be spread on toast, as other jams are, or used as dessert toppings, and many are spicy enough to be served as sauces with roasted meat or game.

The use you plan for the conserve helps determine the fruits that go into it. In this demonstration, for example, three varieties of citrus fruit—lemons, grapefruit and oranges—are combined with pears, raisins and pecans (recipe, page 107). This mixture will yield an excellent spread because the large amount of pectin in the citrus fruit ensures that the long-cooked conserve will set like a jelly as it cools. For a softer, more saucelike conserve to accompany meat, use low-pectin fruit—apricots, for instance, perhaps combined with walnuts and rum.

To extract the most flavor from the fresh fruits, grind them or chop them fine, then combine them and let them stand overnight to release their juices. The next day, the fruits can be cooked in the extracted juices until thick, rich and dark. Dried fruits should receive shorter cooking than fresh fruits so that they retain enough texture to add interest to the conserve. Stir dried fruits into the mixture about halfway through the cooking. Nuts require no cooking and are added at the last minute.

Because dried fruits and nuts are heavier than the chopped fresh fruits, they will tend to sink to the bottom of the pan. When the cooking is completed, allow the mixture to cool and thicken slightly, then stir it so that the ingredients are evenly distributed when the conserve is canned.

1 **Preparing fruits.** With a sharp knife, cut thin slices of unpeeled grapefruit, orange and lemon; remove the seeds and discard them. Quarter, peel and core firm pears.

2 **Grinding fruits.** Fit a food grinder with a coarse disk and fasten the grinder to the edge of a work surface. Place a tray beneath the grinder to catch the fruit and juice. Fill the mouth of the grinder with fruit and grind.

5 **Stirring in raisins.** Pick over raisins and discard any stems. Add the raisins to the simmering conserve mixture, stirring to incorporate them thoroughly.

6 **Adding nuts.** Continue to cook the conserve over low heat for an additional 45 minutes, stirring it frequently. When the conserve has thickened and darkened, stir in shelled pecans and cook for about five minutes more.

3 **Adding sugar.** Tip the ground fruits into a large mixing bowl. Add sugar and stir it into the fruits. Cover the bowl with plastic wrap and leave the mixture overnight at room temperature to let the sugar dissolve and draw out more juices.

4 **Cooking the fruits.** Transfer the fruits to a large, nonreactive pot. Set the pot over medium heat and bring the mixture to a boil. Then reduce the heat and simmer, uncovered, for about 45 minutes, or until the conserve begins to thicken. Stir occasionally with a wooden spoon to prevent the fruits from sticking to the bottom of the pan.

7 **Filling the jars.** Remove the pan from the heat and let the mixture cool briefly. Stir to combine all of the ingredients. Use a ladle and a canning funnel to fill hot, clean jars to within about ½ inch [1 cm.] of the rims. Cover the jars and process (pages 30-31).

Butters and Cheeses: Reductions of Puréed Fruit

If fruits are puréed before they are mixed with sugar and boiled down, they become fruit butters or cheeses—so called because their textures resemble their milk-based namesakes. Creamy fruit butters such as the apple preserve in the top demonstration *(recipe, page 116)* are spread over toast or bread; firm cheeses such as the damson-plum preserve in the bottom demonstration *(recipe, page 118)* are unmolded, sliced and served with cold meats or poultry, or as desserts.

These fruit preserves are economical. You can make them from whole fruits that are too ripe or too bruised for jams and jellies (however, brown spots must be cut away). Or you can use the pulp left over from a fruit syrup *(page 39)*, a pectin stock *(page 45)* or a jelly *(pages 48-49)*—although the results will have less flavor than preserves made from whole fruits. Any fruits will do except citrus, which have too much juice and membrane to be transformed into purées.

To make either butters or cheeses, the puréed fruit is cooked with sugar until enough moisture has evaporated to produce the desired thickness. The different textures of the two preserves will be determined by the amount of sugar you use and the length of the cooking time. To attain the spreadable consistency of a fruit butter, the fruit or pulp is cooked until it is thick, but not stiff, and only enough sugar is added to sweeten the mixture. The dense texture of a fruit cheese is the result of longer cooking and a higher proportion of sugar.

A fruit cheese will keep for at least a year in the refrigerator; it is most easily unmolded if it is packed in jars whose interiors have been coated with tasteless vegetable oil. Butters, because of their lower sugar content, last for only about six months. To keep a butter or cheese longer, process it *(pages 30-31)* in a boiling water bath.

A Smooth Apple Butter

1 Reducing the liquid. To make a butter from whole, firm fruit such as apples, you must cook them in enough water or juice to prevent sticking. If you use juice such as the cider shown here, concentrate its flavor before you add the fruit by boiling it in a heavy, nonreactive pot until it has reduced to half of its original volume.

2 Adding the fruit. Peel, core and chop the apples. When the cider has reduced, lower the heat and add the apples. Simmer the fruit, stirring it frequently with a wooden spoon, for about 25 minutes, or until it is very soft and smooth.

A Firm Cheese from Plum Pulp

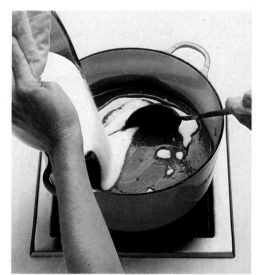

1 Straining. Wash the fruit—in this demonstration, damson plums—and boil it in a little water for 20 minutes, until soft. Purée the fruit to remove pits and skins. Line a strainer with cheesecloth and pour into it the purée and juice. Let the fruit drain overnight; reserve the juice for a jelly *(pages 46-47)*.

2 Cooking. Place the drained purée in a nonreactive pot set over low heat. Simmer the purée, covered, for 45 minutes, until it is thick. Stir in sugar; the sugar can be warmed for 10 minutes in a 300° F. [150° C.] oven, as here, so that it will not lower the temperature of the purée and slow the cooking.

3 **Incorporating sugar.** If you like, add spices to the purées; here, ground cloves, cinnamon and allspice are used. Stir sugar into the mixture *(above, left)*. Continue to simmer the purée — uncovered, and stirring frequently to prevent burning — for about 40 minutes, until it is thick. The butter is ready when a spoon drawn across the base of the pan leaves a trail that is quickly covered by the mixture *(right)*. If the purée has reached the required thickness but still contains lumps of fruit, force it through a sieve to smooth it, then bring it to a simmer again.

4 **Packing the butter.** Remove the pan from the heat and ladle the mixture into hot, clean jars, leaving a ½-inch [1-cm.] headspace. Cover the jars and refrigerate. If you wish to keep the butter for more than six months, process the filled jars in a boiling water bath.

3 **Finishing.** When the sugar dissolves, increase the heat until the purée bubbles gently. Cook for about 50 minutes, stirring frequently to prevent scorching. The cheese is ready when a spoon drawn across the surface of the purée leaves a trough that is slowly filled in by the stiff mixture *(above)*. Remove the pan from the heat.

4 **Unmolding.** Spoon the cheese into oiled canning jars, cover and process. To unmold, dip a knife in hot water and run it around the inside of the jar *(inset)*. Invert the jar and carefully pull it off the cheese. Slice the cheese into rounds; if you like, decorate by stamping the rounds with an aspic cutter.

4

Vinegar and Alcohol Preserves
Steeping Foods in Flavor

A straightforward way to make pickles
Sweet and sour relishes
A geometric display of vegetables
Transforming clear spirits into fruit liqueurs
Making a traditional mincemeat

Red wine vinegar is poured over tiny cucumbers and a variety of fresh and dried flavorings — including small onions, dried chilies, peppercorns, fresh coriander, and sprigs of tarragon and thyme *(page 62)*. The vinegar preserves the cucumbers and imparts its own delicate flavor to the pickles, which should be left to mature in an airtight jar for several weeks before serving.

Vinegar and alcohol both prevent spoilage by penetrating food and by supplanting its natural liquid, acting against microorganisms in the process: Alcohol kills the organisms, vinegar prevents their growth. Alcohol, formed from sugar, is used primarily with fruits to make a diversity of sweet preserves. Vinegar is used with both vegetables and fruits to make a variety of pickles.

The word vinegar comes from the French *vin aigre* — "sour wine." Although wine vinegar has the finest flavor, less expensive cider, malt or distilled grain vinegar can also be used, as long as it is of good quality. To be an effective preservative, the vinegar must contain at least 5 per cent acid, a fact that will be noted on the bottle's label. Because acid reacts with some metals, discoloring food and producing an off taste, all equipment used in pickling must be made of nonreactive materials.

The simplest pickles are made by packing raw vegetables or fruits into jars and immersing them in vinegar. To add more tastes — and to soften the vinegar's tartness — herbs, spices and sugar are often included. For a distinctive complexity of flavor, vinegar can be combined with other ingredients such as mustard and made into a thick, rich sauce for cooked pickles. Vinegar also serves in the making of long-lasting ketchups — the concentrated fluids of vegetables, fruits or nuts, used to flavor other foods during cooking.

Like vinegar, alcohol flavors as well as preserves. Beer and wine are too low in alcohol to act alone, but any distilled spirit, such as brandy, will work effectively. Often the resulting liqueur is as much prized as the food it preserves. On page 70, for example, cherries are bottled with *eau de vie,* which becomes a cherry-flavored homemade liqueur.

Mincemeat is another classic preserve in which alcohol is used. The traditional version *(pages 72-73)* is made with cooked beef tongue and brisket, grated suet, dried and candied fruits, sugar, spices and brandy. High proportions of both sugar and alcohol prevent decay.

In any vinegar or alcohol preserve, the amount of vinegar or spirit needed will vary according to the nature of the food. For safety's sake, the amount specified in a recipe should never be reduced. To keep the preserving liquid from evaporating, any food in alcohol or vinegar should be tightly covered and — if it is not processed in a water bath — refrigerated.

A Two-Stage Technique for Pickling

The simplest method of pickling vegetables in vinegar is short brining, a two-stage process. First, the vegetables are sprinkled with salt and left for about 24 hours; the salt draws out juices that would dilute the vinegar. (Use coarse pickling or kosher salt: Table and sea salt contain minerals that could cloud the pickles.) Then the vegetables are rinsed, drained, packed into jars, covered with vinegar and aged to absorb the vinegar.

If vinegar is brought to a boil before it is poured over the vegetables, it penetrates them quickly and softens them slightly. Hot vinegar is chosen for vegetables, such as beets, that are large and firm enough to retain their textures (recipe, page 128). Cold vinegar is preferred for smaller, more delicate vegetables—the small cucumbers shown here, for example (recipe, page 130)—to ensure that they remain crisp.

In either case, use vinegar that contains at least 5 per cent acetic acid. Usually the choice of vinegar depends on the flavor you want, but if pale vegetables such as onions or cauliflowers are being pickled, color is a factor: Only white wine or distilled vinegars do not alter the vegetables' color.

For flavor, include aromatics such as onions and garlic, and pack herbs and spices into the jars. Thyme, tarragon and bay leaves are popular pickling herbs, and dill is classic. Pickling spices include coriander, cloves and mustard seeds. Use whole spices; ground spices or the crumbled mixture in commercial "pickling spices" cloud the brine. To make vinegar unencumbered by seeds, leaves or stems, boil the flavorings in the vinegar, let it cool, then strain it.

Cucumbers are so often selected for pickling that the process is almost synonymous with them. Choose them with care: Cucumbers sold in supermarkets usually are waxed to keep them fresh; these are unsuitable for pickling because the wax prevents their absorbing vinegar. Instead, use the fresh, unwaxed cucumbers sold at vegetable stands—or ripened in your garden—during the summer. The tiny pickling cucumbers seen in this demonstration are available early in the season.

1 **Cleaning cucumbers.** Pick over the cucumbers, removing any blossoms attached to them and discarding any cucumbers that are soft or bruised. Wash the cucumbers briefly in cold water, then lightly rub each one with a cloth to remove its fine, downy spikes.

2 **Salting the cucumbers.** Sprinkle a thin layer of coarse salt in the bottom of a nonreactive bowl and place a layer of cucumbers on top. Then fill the bowl with alternating layers of salt and cucumbers, finishing with a layer of salt. Cover the bowl and let the vegetables steep at room temperature for 24 hours.

5 **Filling the jars.** Sterilize glass jars by boiling them for 10 minutes in water—clamp-top preserving jars (page 10) are used here. Fill the jars with the cucumbers and distribute the flavorings among them; in this case, the flavorings are peeled garlic cloves and small onions, coriander seeds, whole cloves, peppercorns, small sprigs of tarragon and thyme, one or two dried chilies, and bay leaves. Leave about ½ inch [1 cm.] headspace at the top of the jars.

3 **Removing the salt.** Fill a large bowl with water and a little vinegar; the vinegar will help draw excess salt from the cucumbers. Transfer the cucumbers to the bowl in handfuls *(above)*, then swirl them in the vinegared water to wash off excess salt.

4 **Drying the vegetables.** Spread a large clean cloth over a work surface and place a layer of the washed cucumbers on one half of the cloth. Fold over the other half of the cloth and pat the cucumbers dry. Use another cloth to dry the remaining cucumbers if the first cloth becomes very damp.

6 **Adding vinegar.** Pour vinegar — in this instance, red wine vinegar — over the cucumbers, covering them completely. Cover the jars with their clamp tops, but do not process the jars as described on pages 30-31: The heat would rob the pickles of their crispness. Instead, store the pickles in the refrigerator; they will be pickled and ready to eat in three to four weeks and will keep in excellent condition for six months.

7 **Serving the pickles.** Unclamp a jar and use wooden serving tongs to lift out as many pickles as you need. Leave the remaining pickles in the vinegar, re-cover the jar and refrigerate. Cut the pickles into thin slices to add to salads or sauces, or serve them whole or chopped as an accompaniment to cold meats.

Ketchup — An Essence Preserved by Vinegar

By extracting the juice of fruits, nuts, fish or vegetables, reducing it to its essence and mixing it with vinegar to preserve it, you can make a thin but intensely flavored ketchup. The word itself derives from that for a spicy Malaysian fish-based juice, and the mixture is used in much the same way as in the Orient.

Unlike a thick ketchup that is served as a condiment with meats and vegetables, this concentrated liquid is usually added to food during cooking: A few drops will add a distinctive piquancy to broiled or roasted meats and fish, as well as to soups, stews and gravies. Mushroom juice, its flavor sharpened by the addition of salt anchovies, shallots, fresh ginger and spices, forms the ketchup shown at right *(recipe, page 123);* lobsters, lemons, green walnuts, blackberries and apples are among the other foods that serve as ketchup bases *(recipes, pages 121-124).*

To draw out most of their liquid, the vegetables, fruits or nuts are first salted and left to steep for about nine days. They are then cooked by gentle, lengthy simmering, along with the liquid, to extract the remaining juice. Next, the pulp is strained out, and the liquid is combined with flavorings and vinegar, and cooked again to reduce it and concentrate its flavor. A final straining produces a clear brown ketchup ready for canning. When processed *(pages 30-31),* the ketchup can be stored for up to one year.

1 Preparing mushrooms. Trim the stems of the mushrooms and wipe each mushroom with a damp cloth. Break up the mushrooms, place them in a bowl and sprinkle pure, coarse salt over them — kosher salt is used here.

2 Mixing in the salt. With your hands, mix the salt and mushrooms thoroughly. Cover the bowl with foil or plastic wrap, and refrigerate it. Stir the mushrooms daily for eight to 10 days to make sure that all of the mushroom pieces remain evenly coated.

6 Adding vinegar. Soak, rinse and fillet salt anchovies *(page 83);* assemble mace blades, whole cloves, peppercorns, shallots and sliced ginger. Measure the mushroom juice and pour it into an enameled or stainless-steel pan. Add vinegar — 1 cup [¼ liter] for every 2 cups [½ liter] of juice — and stir in the other ingredients.

7 Cooking and straining. Bring the mixture to a boil over medium heat. Then simmer it uncovered until the liquid is reduced by half. Remove the pan from the heat. Fold a piece of muslin or cheesecloth into quarters and set it in a strainer placed over a large bowl. Pour the ketchup through the cloth *(above)* to yield a dark, clear liquid.

3 **Filling the crock.** When the mushrooms have turned dark brown and are soft enough to bend easily, transfer them with their liquid to a large, heat-resistant crock that can be fitted inside a deep pot. Then cover the crock tightly with foil.

4 **Cooking the mushrooms.** To ensure gentle cooking, put a wire rack in a pot and set the crock on the rack. Add warm water to the pot, up to the top level of the mushrooms in the crock. Cover the pot. Simmer for three hours. Add boiling water as necessary to maintain the original water level.

5 **Straining.** Take the pot off the heat, lift out the crock and remove the foil. Set a strainer over a bowl and ladle in some of the crock's contents. With a wooden pestle, press the pulp firmly to extract as much of the mushroom liquid as possible; discard the pulp. Repeat this process until all of the mushroom juice has been extracted.

8 **Filling and processing.** Ladle the hot ketchup into warm, clean canning jars — here, 1½-cup [375-ml.] jars. Leave a headspace of ½ inch [1 cm.]. Cover the jars and process them in a boiling water bath.

Combining Vegetable Relishes

A Medley in a Mustard Sauce

Mixed vegetables may be pickled in two ways: You can cook them together to make a relatively homogeneous relish, or you can cook them separately so that they retain as much of their individual characters as possible.

Relishes are made by salting vegetables to extract their excess liquid, then simmering the ingredients in a vinegar-based sauce. Cooking softens the vegetables somewhat and also imbues them with the sauce's flavor; the sauce, in fact, becomes as much a part of the pickled product as the vegetables. In the demonstration at right, the sauce includes vinegar, dry mustard, sugar and spices; at the end of cooking, the sauce is thickened with cornstarch. The result is the pungent mustard relish known as piccalilli *(recipe, page 144)*.

By adjusting the proportions of mustard and spices to vinegar, you can make the relish as hot or as mild as you like. You can vary the cooking time, too, depending on how crisp or tender you want the vegetables to be.

Relishes of this type are distinguished by unity of color and flavor. If you want to produce mixed pickles that are differentiated rather than unified, first blanch each of the vegetables in boiling salted water *(opposite, bottom)*. Blanching, like salting, leaches out excess liquid, and it tenderizes the vegetables. By blanching the different types separately, you will ensure that each vegetable receives no more cooking than it needs: Its texture, color and flavor are minimally altered.

To provide visual emphasis for such mixed pickles, arrange the vegetables in distinct layers in their canning jars, and cover them with a clear mixture of vinegar and olive oil, which displays their contrasting colors and shapes.

1 Salting the vegetables. Wash the vegetables — cauliflower, cucumbers, zucchini, green beans, small boiling onions, red peppers, carrots and celery are used here. Cut them into bite-sized pieces. Put them in a bowl and mix in pure, coarse salt; cover the bowl and allow the vegetables to drain overnight.

2 Rinsing the vegetables. The next day, empty the vegetables into a deep bowl or basin of water, and rinse them thoroughly to remove the salt. Drain the vegetables in a large colander.

5 Thickening the sauce. In a small bowl, mix cornstarch with a little cold vinegar so that it will blend evenly into the sauce. Place the pan of sauce back on the heat. Stirring constantly, pour in the cornstarch mixture and cook the sauce briefly at a simmer for about five minutes, or until it thickens.

6 Adding the sauce. Take the pan from the heat. Fill each jar with sauce, leaving a ½-inch [1-cm.] headspace. Make sure that the vegetables are completely submerged. Cover the jars and process *(pages 30-31)* in a boiling water bath. To allow the relish to mature in flavor, store it for at least one month before serving.

3 **Preparing the sauce.** Put sugar, turmeric, dry mustard and ground ginger in a nonreactive pan. Pour in a little vinegar. Over medium heat, stir the ingredients with a wooden spoon to dissolve them, then add the remaining vinegar *(above, left)* and bring the sauce to a boil. Add all of the vegetables *(right)* and reduce the heat. Stirring with a wooden spoon, simmer the mixture, uncovered, for about 15 minutes, or until the vegetables are tender.

4 **Filling the jars.** Remove the pan from the heat. Using a perforated spoon to drain off the sauce, transfer all of the vegetables into warm, clean canning jars. Fill each of the jars to within ½ inch [1 cm.] of its rim.

A Clear Dressing for a Decorative Pack

1 **Blanching the vegetables.** Wash and prepare vegetables — in this case, cauliflower florets, green beans, boiling onions, quartered artichoke bottoms, sliced and julienne carrots, whole and halved Brussels sprouts and mushrooms, and halved, seeded, deribbed and sliced red and green peppers. Blanch each vegetable separately for a few minutes in boiling salted water: The vegetables should remain crisp.

2 **Packing the jars.** Fill clean canning jars with vegetables, leaving a ½-inch [1-cm.] headspace. You may include a few drained pickled cucumbers *(page 63).* Boil olive oil and distilled or white wine vinegar together. Cool this mixture slightly, then pour it over the vegetables. Cover the jars, and process in a boiling water bath for 10 minutes. Store for at least a month before serving.

Chutneys and Other Fruit Pickles

There are two radically different types of fruit pickles, both of which make excellent relishes for meat. One type uses long cooking with vinegar and sugar to reduce the fruit to a pulp that is concentrated in flavor and rich in color *(top demonstration)*. These pickles, a staple in India, are usually called chutney, after the Hindi word for "to taste." The other type is similar to the vegetable pickles on pages 66-67: The fruit is cooked only long enough to tenderize it, and it is canned in a spicy vinegar sauce *(bottom demonstration)*.

Chutney, which can be made from almost any mixture of fruits, usually includes a few aromatic vegetables *(recipes, pages 136-139)*—the combination of apples and green tomatoes shown here is typical—and is also a clever way to use up tomatoes that are still green at the end of their growing season.

Preservation of chutney is assured by its blend of sugars—both natural fruit sugar and refined sugar—and the vinegar's acid. Ordinary white sugar and any commercial vinegar containing at least 5 per cent acid would produce adequate results, but brown sugar and malt or red wine vinegar are most often used. They contribute more flavor and give chutney its characteristic dark color.

Briefly cooked pickles can be made with a single fruit or, like chutney, with any combination of fruits *(recipe, page 139)*. Here, eight different kinds of fruits are cooked in a sugar syrup until just tender, then covered with a smooth, thick sauce made from sugar, vinegar and dry mustard. As with chutney, the blending of sugar and vinegar preserves the fruits and, together with the mustard, gives the pickles a sweet and sour flavor. To avoid darkening the fruits, use white sugar and distilled or white wine vinegar. You can also add whole or ground spices—cloves, allspice or nutmeg—or use flavored vinegar *(recipes, pages 120-121)*.

Canned, sealed and stored in a cool, dark place, all fruit pickles will continue to develop and improve in flavor. Let the pickles mature for at least a month before serving; they will keep for a year.

Slow Cooking for Mellow Flavor

1 **Preparing the ingredients.** Chop fruits and vegetables — in this case, peeled and cored apples, and peeled onions and garlic. Mix the fruits and vegetables with dried currants, raisins, brown sugar and a little salt in a large, nonreactive pot. Wash green tomatoes, cut out their cores and chop them into even-sized pieces; add them to the pot.

A Sweet Mixture in a Pungent Sauce

1 **Preparing fruits.** Wash and prepare fruits — in this demonstration, grapes, sliced lemons and limes, halved and pitted plums and apricots, peeled and cubed honeydew melon and pineapple, and quartered nectarines. In a nonreactive pot, simmer sugar and water briefly together. Add the fruits.

2 **Cooking.** Stir the mixture with a wooden spoon, place the pot over low heat, and simmer the fruits for 10 to 15 minutes, until they are barely tender. Turn off the heat and ladle the fruit and syrup into a large bowl to cool.

2 **Adding vinegar.** Fill a spice bag *(page 36)* with peeled, crushed fresh ginger and split, seeded dried chilies. Add the bag to the pot and set over low heat. Pour in enough vinegar to cover the fruits and vegetables — red wine vinegar is used here. Stir the mixture with a wooden spoon.

3 **Cooking the chutney.** Stirring occasionally, cook the mixture for two hours or so, until it has thickened and reduced to about half its original volume. As the mixture thickens, stir it more frequently to prevent it from sticking. Turn off the heat and remove the spice bag.

4 **Canning the chutney.** Ladle the chutney into warm, clean jars to within ½ inch [1 cm.] of their rims. Cover, and process the jars *(pages 30-31)* in a boiling water bath.

3 **Making the sauce.** In a nonreactive pan, stir sugar into vinegar — here, white wine vinegar — and simmer for about 15 minutes. Cool. Pour dry mustard into a bowl. Stir in the syrup a little at a time *(above),* then let the sauce thicken at room temperature for about one hour.

4 **Saucing the fruits.** Pour the thick vinegar sauce over the fruits and syrup in the bowl *(above).* Using a wooden spoon, mix thoroughly to combine the sauce with the syrup and fruits.

5 **Filling the jars.** Fill warm, clean canning jars with fruits and enough liquid to cover them completely. Leave a headspace of about ½ inch [1 cm.]. Cover, and process the jars in a boiling water bath.

A Reciprocal Union of Fruit and Alcohol

Alcohol is one of the most effective preservatives—if it is present in sufficient concentration. Distilled spirits such as whiskey, brandy, gin, vodka and rum are at least 80 proof. That is, they contain at least 40 per cent alcohol, enough to kill most spoilage microorganisms. Used as a preserving marinade for fruit, these spirits provide a double dividend: The fruit absorbs the alcohol, becoming rich and strong-tasting; the alcohol, flavored by the fruit, makes a fragrant liqueur.

For clarity of flavor, preserve only a single type of fruit; cherries are used here. The fruit should be barely ripe; preserves made with mature fruit will become too soft. To counteract the alcohol's tendency to shrink fruit, mix it with sugar, made into a syrup *(Step 3, below)* for smooth blending. Allow about 2 cups [½ liter] of alcohol and ½ cup [125 ml.] of sugar for each pound [½ kg.] of fruit. Do not process the mixture as described on pages 30-31; the heat would damage the texture of the fruit. Refrigerated, the preserves will keep indefinitely.

1 Preparing cherries. Wash cherries in cold water and drain them in a colander. Cut off only about half of each cherry stem. The part that remains will provide a convenient handle for eating the preserved fruit.

2 Filling the jars. Place the cherries in sterile jars, filling the jars to within ½ inch [1 cm.] of the rims. If you like, add a few whole cloves and a small piece of stick cinnamon to each jar for flavor.

3 Making a syrup. Put sugar into a pan with ¼ cup [50 ml.] of water for every cup [¼ liter] of sugar. Stir over low heat until the sugar dissolves, then put a candy thermometer into the syrup and bring to a boil, without stirring. Cook until the syrup reaches 250° F. [130° C.]. Cool slightly and stir in spirits — here, *eau de vie*, a colorless French brandy.

4 Covering the fruit. Let the syrup cool, then pour it into the jars until the cherries are just covered. Close the jars and store them in the refrigerator.

5 Serving. Marinate the fruit for at least two weeks: The flavors of the spirit and the cherries will improve with time. To serve as a dessert, put a few cherries into a small glass for each diner and cover the fruit with the spirit. Eat the cherries, then drink the liqueur.

Progressive Melding of Just-ripe Fruits

Among the most elegant of fruit-and-alcohol preserves are those in which a fruit syrup—fruit juice mixed with sugar and spirits—serves as a marinade for an assortment of fruits *(recipe, page 150)*. Long steeping mellows the mixture into a rich sauce for ice cream or cake.

The basic juice may be made from almost any soft fruit; strawberries are used here, but raspberries or peaches would also serve. The juice is enriched with brandy, but other spirits may be substituted; in Germany, a preserve of this type is traditionally made with rum and is called *rumtopf*—literally, "rum pot."

Except for apples, pears and bananas, whose textures suffer from marination, almost any fruit can be added. For the best results, add firm, ripe fruits serially, as they come into season. Of the fruits used in this demonstration, begin marinating strawberries in June; other fruits are added during the summer and fall. Each time you add fruit, you must also add sugar and alcohol *(Steps 4 and 5, below, right)* for effective preserving.

1 **Preparing juice.** Place the first soft fruit of the season — strawberries are shown — in a nonreactive pot set over medium heat. With a pestle or a potato masher *(above)*, crush the fruit and simmer it for approximately 10 minutes, until it is soft. Set a strainer lined with a double layer of cheesecloth over a bowl and pour the fruit into it.

2 **Sweetening the juice.** With a wooden spoon or pestle, press on the fruit pulp to extract as much juice as possible; discard the fruit pulp. Measure the hot juice and stir sugar into the juice, allowing the same volume of sugar as liquid. Stir until the sugar dissolves.

3 **Ripening the syrup.** Pour the juice into a clean ceramic crock. Add a spice bag *(page 36)*; in this case, the bag is filled with stick cinnamon, whole cloves, grated lemon peel, allspice and sliced fresh ginger. Pour in enough brandy to equal the volume of the juice. Cover the crock, set it in a cool place and let the mixture steep for one week.

4 **Adding fruit.** Add ripe fruit — in this instance, washed, hulled strawberries. Stir in the same amount of sugar as you have fruit and add one quarter as much brandy. Continue adding trimmed fruit, sugar and brandy in these proportions as fruits come into season, allowing four to seven days between any additions.

5 **Marinating.** Add the last fruit, sugar and brandy. Here, grapes, raspberries and pineapple are added to a mixture that includes cherries, peaches and plums as well as strawberries. Cover the crock, tie cheesecloth tightly over the top, and put in a cool place for two to three months before serving.

A Rich Amalgam of Fruit and Meat

Alcohol forms an exciting alliance with a wide range of foods in addition to fruits. In this demonstration, for example, brandy and sherry are used to preserve a traditional mincemeat pie filling *(recipe, page 159)* made of cooked beef, jam, sugar, dried fruits and spices. The mixture is enriched with suet (the white fat that surrounds beef kidney).

The white sugar and the fruits' natural sugars will help to preserve the mixture. However, because of the perishable nature of meat and the high proportion of meat used in the mincemeat mixture—approximately 7 pounds [3½ kg.] are used here—you will need at least 1½ cups [375 ml.] of alcohol per pound [½ kg.] of mincemeat mixture.

After the mincemeat is assembled, it is placed in a large, airtight crock and left to mature in a cool, dry place for at least a month so that the ingredients have time to become thoroughly impregnated with alcohol and sugar. Once a week, the mixture is stirred to ensure even absorption. The ingredients should always be submerged in liquid to keep the mincemeat moist and to prevent bacterial growth; add brandy and sherry as needed to compensate for absorption and evaporation.

If it is tightly covered and refrigerated, mincemeat will keep for six months. For longer keeping—up to a year—process matured mincemeat in a pressure canner as specified on pages 30-31.

1 Cooking the meat. Put the meat—here, a fresh beef tongue and a 3-pound [1½-kg.] beef rump roast—into a large pan of cold salted water. Bring the water to a boil and remove any scum with a spoon. Cover the pan and reduce the heat. Simmer the meat for about two and one half hours, or until a skewer will pierce it easily.

2 Peeling the tongue. Turn off the heat, let the meat cool slightly in its cooking liquid, then lift it out with a large fork. Cut the rump into small cubes and put them in a bowl. Remove the neck bones from the base of the tongue with your fingers. Slit the tough skin open and peel it off *(above)*. Cut the tongue into cubes.

5 Combining the ingredients. Shred fresh lemon and orange peel, and chop candied citron. Put them into a bowl with sugar and dried fruits—currants and seedless dark and golden raisins are shown. Add the suet, meat and flavorings—strawberry jam, salt, and ground cinnamon, nutmeg, allspice, mace and cloves are used here. Pour in sherry and mix with your hands. Add enough brandy to make a loose mixture.

3 **Grinding the meat.** Fit a coarse disk into a food grinder and set a tray underneath it. Feed the meat cubes into the grinder a few at a time, pressing the cubes lightly with a pestle or your fingers. Alternatively, chop the meat cubes fine with two knives, as demonstrated on page 25.

4 **Grating suet.** Pull off the thin outer membrane from a piece of suet. Break the suet apart and remove any internal membrane. Shred the suet on the small holes of a box grater, as shown here, or chop it very fine with a knife.

6 **Storing.** Rinse a stoneware crock and its lid in boiling water. Tip in the mincemeat and cover *(inset)*. Put in a cool, dry place (40° to 60° F. [4° to 15° C.]) or in the refrigerator for at least a month, stirring the mixture once a week and adding more brandy or sherry to cover the fruit. Use the mincemeat within six months, or transfer it to clean jars *(right)* and process at 10 pounds of pressure [70 kPa] for 20 minutes.

5

Fat-sealing, Drying and Brining

Tactics that Focus on Moisture

Rillettes: tender shreds of meat
Concentrating the fragrance of herbs
How to use a dehydrator
Making caviar at home
Fermenting sauerkraut and dill pickles
Cold smoke to preserve fish

Microorganisms need moisture. Without it, they cannot grow and therefore cannot spoil food. The preserving techniques demonstrated in this chapter all aim to deprive spoilage organisms of moisture—by sealing food in fat, by adding large amounts of salt to food, or by drying food in warm, circulating air or smoke.

A seal of fat on meats excludes moisture and airborne bacteria, but does not affect the moisture or microorganisms present in the food. However, if the food is first well cooked, most microorganisms in it will be killed and enough moisture will evaporate so that the food will keep for several months; because cooking does not destroy all microorganisms, the food must be refrigerated. A sumptuous example of this preserving method is French rillettes, pork—sometimes mixed with other meat—cooked till tender, shredded and sealed with its own fat *(pages 76-77)*.

Air-drying—simply hanging food up in a dry place—would seem to be the easiest way to rid food of moisture. In practice, however, the method is problematic for all but small items or those that are relatively low in moisture—chilies, for example, or bunches of herbs *(pages 78-79)*: These dry so quickly in still air that they suffer no deterioration during the process. Larger or moister foods would spoil if treated this way, but they may be dried in a very slow oven or with an electrically powered food dehydrator. The soup mix of dried vegetables shown opposite is an example of the results obtainable from careful home drying. A related technique is smoking *(page 90)*. In principle no different from oven-drying, smoking gives meats or fish a subtle, woody aroma.

Salt draws moisture from food cells, forming a brine in which harmful microorganisms cannot grow. For effective preservation, the salt should penetrate the food thoroughly and evenly. Relatively thin, flat foods such as anchovies are easily preserved by the method known as dry-salting: The food is layered or coated with salt, which gradually forms a brine with the food's juices. However, the slowness of dry-salting makes it unsuitable for fragile foods such as the fish roe used to make caviar; these can withstand only the briefest salting. It also works too slowly for large pieces of meat, which can spoil before they are thoroughly impregnated with salt. Such foods should be preserved in an already prepared salt brine, which works more quickly than salt alone.

Corn kernels, peas, and sliced carrots and green beans—their colors and flavors intensified by slow dehydration—are spooned into jars for storage. Tight-fitting lids with rubber seals will keep moisture out until the dried vegetables are reconstituted in the liquid of a soup or stew.

A Protective Shield of Fat

Fat can provide a seal that excludes air—and the microorganisms it bears—from meat. If the meat has first been cooked for a prolonged period to reduce its moisture content and discourage the growth of bacteria, a shield of fat will provide effective short-term preservation: Sealed meat—reduced to fine, tender shreds by its lengthy cooking—will keep in a refrigerator for two months.

This preserving method is best suited to meats that have a natural abundance of good-flavored fat—especially goose, duck and pork. Leaner meat such as rabbit could also be used, but it should be mixed with pork for the necessary protective seal. All such preserves, known by the French term rillettes (*right; recipes, pages 160-162*), can be spread on toast or bread to produce excellent hors d'oeuvre.

The rillettes in this demonstration are made with a combination of goose and pork. The meat receives more than eight hours of cooking and becomes so tender that the goose flesh literally falls off the bones. After cooking, the rendered fat is strained from the meat and reserved. Then the cooled meat is mashed; in order to locate and remove any small bones, it is best to do this mashing with your hands, but if the rillettes were made with boneless meat, you could use a fork. The mashed meat is mixed with some of the reserved fat to create the rillette paste, which is packed into small crocks or jars and covered with the remaining fat before being refrigerated.

Many preserves like these—known in England as potted meats, from the small, earthenware containers in which they are stored—are cooked more briefly than rillettes. Large pieces of meat—tongue, for example (*recipes, pages 160-162*)—are cooked until just tender, pounded to a paste, then covered with melted fat. To draw out its moisture, the meat is often salted before cooking (*page 89*); even so, such preserves are moister than rillettes and have a shorter storage time. They are best eaten within a few weeks.

1 **Cooking the meats.** In a heavy casserole, put diced lean and fatty pork. Add finely chopped onion and carrot, salt, and mixed dried herbs. Cut a goose into small pieces and arrange them on top of the other ingredients (*above*). To prevent the meats from sticking before any fat has been rendered, add water to a depth of about ½ inch [1 cm.]. Cover the casserole and put it over medium heat; after the water comes to a boil, reduce the heat to very low, and cook the meats until the water evaporates and the goose flesh falls off the bones — about eight hours.

5 **Warming the meat.** Return the mashed meat to the casserole. Add the fat left in the bowl and any fat on the meat tray. Over low heat, stir continuously with a wooden spoon to combine the meat and fat thoroughly. When the mixture is heated through and well blended — about 15 minutes — remove the casserole from the heat.

6 **Packing the rillettes.** Sterilize small containers (*page 10*); glazed earthenware pots are used here. Fill each one to within ½ inch [1 cm.] of its rim with the warm meat mixture. To get rid of air pockets, pack the meat firmly with a spoon (*above*). As you work, stir the meat in the casserole occasionally to keep the fat distributed.

2 **Straining the meat.** Take the casserole off the heat. Set a colander over a bowl and pour in the contents of the casserole. Leave the liquid fat in the bowl for about 10 minutes so that meat particles and other residue can settle to the bottom. Meanwhile, set the colander of meat over a large tray.

3 **Transferring the liquid fat.** With a ladle, transfer all of the clear fat from the mixing bowl to a saucepan. Set it aside: You will need it later to seal the pots. Leave the residue in the bowl: It will be mixed into the rillette paste.

4 **Mashing the meat.** With your fingers, pick out and discard the bones from the goose meat. Do not discard the skin: It will lend its own texture and flavor to the rillettes. Squeeze the goose and pork meat together into a paste (above), discarding any small bones that you find.

7 **Sealing.** Pour a ½-inch [1-cm.] layer of the reserved clear fat over the meat. Cover the pots and, along with the remaining fat, refrigerate overnight. The fat will shrink from the sides of the pots. To seal, melt the reserved fat and top the chilled pots with it. When this fat sets, cover the pots and refrigerate.

8 **Serving the rillettes.** The rillettes will keep in the refrigerator for up to two months. Serve them cold, straight from storage, or at room temperature. Spoon portions of the rillettes —complete with some of the protective layer of fat —direct from the pot. Once the seal of fat is broken, the meat should be consumed within a week or two. Serve it with toast or crusty bread.

A Spectrum of Dried Flavorings

The simplest and most straightforward way to preserve herbs and other plant flavorings is to hang them in a dry, shady place. By eliminating the moisture that organisms need for growth, drying prevents spoilage. At the same time, drying concentrates the plants' essential oils and provides a store of flavor for future cooking. For example, strips of citrus peel, cut from lemons, tangerines or oranges *(right)*, add tang to stews and soups as well as to desserts. Dried shallots, ginger or chilies *(far right)* enhance meats, chutneys and pickles. Essential to most good cooking are herbs *(right, bottom)*, picked and dried at their prime.

Some herbs are more suited to drying than others. Among those that best retain their natural flavors are thyme, rosemary, marjoram, savory, sage, wild fennel and oregano. Dried tarragon and mint also have appeal, although their flavor is very different from that of their fresh counterparts. Chervil, parsley and chives, however, have little flavor after drying and should be used fresh.

Although herbs can be harvested and dried throughout their growing season, the ideal time for gathering thyme, marjoram and oregano is when they come into flower; their blossoms are even more fragrant than the leaves.

Plant flavorings of all types are dried simply by hanging them in a spot that is fairly dry and warm—an airy corner of the kitchen, a garden shed or the attic. When the flavorings are completely dry to the touch, store them in airtight containers to keep them free of dust and to help retain their fragrance. Before storing, dried herbs can be ground and sieved to eliminate woody stems.

Store all flavorings in a dry place that is not exposed to direct sunlight; the sun's rays would eventually rob them of their color and taste.

Threading Citrus Peels

1 Removing the citrus peel. Select large, unblemished oranges. Wash the oranges. Peel each one by rotating it slowly in your hand and paring away the thin outside layer of skin with a small, sharp knife or a vegetable peeler. Be careful not to cut so deep that you include any of the bitter, white pith.

2 Stringing the peel. Thread a large needle—a trussing needle, in this case—with string and push the needle through each strip of peel *(above)*. Leave about 1 inch [2½ cm.] of string between strips so that air can circulate freely around them.

Blending a Mixture of Herbs

1 Tying up the herbs. Loosely bunch the herbs—thyme is shown—to allow the air to circulate. Wind a piece of raffia *(above)* or string around their stems, knot it securely, and make a loop with the free ends. Hang the herbs to dry for two to three weeks, until a leaf crumbles when touched.

2 Crumbling herbs. Working over a tray and using gloves to protect your hands from the scratchy stems, rub the herbs between your palms. Save the stems to flavor vinegars. If you like, combine herbs; here, thyme, marjoram and savory produce a mixture for flavoring soups and stews.

Hanging a String of Chilies

3 Drying the peel. Hang the strings of peel and leave them until they are completely dry — three to four days. To store the peel, pull the strips from the string and put them into an airtight container *(inset)*; their fragrance will last for a year or longer.

1 Threading chili peppers. Select firm hot chilies — red, green or, as shown here, a mixture. Thread a trussing needle with string and knot one end. Push the needle through the stem of each chili and string the chilies alternately left and right. Make a loop on the top end of the string and knot it *(above)* so that it can be used to hang the chilies.

3 Grinding the herbs. Use a food processor — as here — or a blender to grind the herbs to a fine powder; this should take less than a minute. Tip the ground herbs into a large bowl.

4 Sieving dried herbs. To eliminate woody pieces, place a handful of the herbs in a fine-meshed sieve set over a bowl. Rub the herbs through the sieve with one gloved hand. Discard the fragments that are left in the sieve. Sieve the remaining herbs and store them in airtight containers.

2 Drying the chilies. Hang the chilies for at least two weeks, until they shrivel and feel dry. Pick them off the string as required *(above)*. They will keep their flavor for a year or two. Wash your hands carefully after handling fresh or dried chilies: Their oils can irritate eyes or sensitive skin.

Drying Dense Foods in Warm, Moving Air

Vegetables, fruits and meats, too dense to be dried by the simple method used for herbs *(pages 78-79)*, will yield their moisture when they are exposed to warm, dry, circulating air for long periods.

The method began in regions blessed with a very clean, dry atmosphere and an unbroken succession of bright summer days. Although such conditions are unavailable to most people, they may be recreated indoors with no great difficulty. One tactic is to arrange the prepared food on racks in a slow oven; the oven door is kept open and an electric fan placed in front of it to blow air around the food. Or use an electric dehydrator. These devices, available at health-food stores, maintain the temperature and air circulation needed for drying.

As indicated by the top demonstration at right, this approach to preserving will work with almost any vegetable; the only exceptions are the leaf vegetables, which would wilt. Before drying, the vegetables should be trimmed and cut into small pieces to expose as much surface area as possible; all except green peppers and onions should be blanched as for freezing *(guide, pages 14-15)*. Treated this way, most firm vegetables will shrivel, dry and become concentrated in flavor within two to eight hours. The vegetables are so dry that they must be rehydrated in liquid such as soup broth to be palatable.

Fruits can be dried in the same way as vegetables, in about the same time. Or, for a ready-made snack, you can make fruit leather *(right, bottom)* by puréeing the fruit, then drying it in the oven or dehydrator. To preserve their color, pale fruits should be blanched before puréeing. The purées may be mixed with spices such as cinnamon, cloves or allspice and sweetened with honey or corn syrup; sugar would make the leather grainy.

The oven or dehydrator also can be used to dry paper-thin slices of meat—beef and venison are preferred for their assertive tastes—to make chewy jerky *(opposite, bottom)*. Use lean meat; fat may become rancid during drying. For extra flavor, the meats can be salted or brined *(pages 88-89)* or, as here, marinated before drying. The jerky may be eaten as is or rehydrated in a stew.

Dehydrating a Mélange of Vegetables

1 **Preparing vegetables.** Trim green beans, cut them into ½-inch [1-cm.] pieces, and put them in a bowl. Peel small carrots, slice them into pieces ½ inch thick, and place them in a separate bowl. Shell peas and put them in a third bowl. Husk corn. Blanch and drain each vegetable separately.

2 **Drying vegetables.** Preheat the oven or, as here, a dehydrator, to 140° F. [60° C.]. Spread each prepared vegetable in a single layer on a separate tray, removing the kernels of corn from the cobs. Push all the trays in and close the dehydrator.

Baking a Purée to Create Fruit Leather

1 **Spreading purée.** Peel, pit and cut up fruit; nectarines and bananas are used here. Put the fruit with a little water into a heavy pot and simmer it for five minutes. Purée the fruit in a food processor or force it through a food mill or sieve. Let it cool. Spread the purée ⅛ inch [3 mm.] thick in a jelly-roll pan that has been sprinkled with water and lined with plastic wrap.

2 **Drying the purée.** Put the pan in a preheated 140° F. [60° C.] oven; leave the oven door open. Place an electric fan in front of the oven so that the moving air is directed at the fruit. Dry the purée for seven or eight hours, until it darkens and feels tacky—not moist. Roll up the fruit leather in the plastic wrap and store in a cool, dry place.

3 **Checking for doneness.** After about two hours of drying, begin checking each vegetable. Halfway through the process *(above, left)*, the vegetables will be shriveled, but still pliable. When drying is complete *(right)*, the vegetables will be dark, shriveled and brittle. Remove each vegetable when it reaches this point.

4 **Pasteurizing vegetables.** Line a jelly-roll pan with aluminum foil, pleated to form a compartment for each vegetable. As each tray is removed from the dehydrator, empty it into a compartment. In order to deactivate any enzymes that might be left in the vegetables, put the pan in a 170° F. [75° C.] oven for 15 minutes.

5 **Sorting vegetables.** Remove the vegetables from the oven and let them cool. Spoon the vegetables into small canning jars, arranging the vegetables in separate layers for effect. Close the jars and store them in a cool, dark place.

Turning Meat Strips into Jerky

1 **Marinating the meat.** For easy, uniform slicing, put the meat — in this case, flank steak — in the freezer until it is very firm. Cut the meat into strips ¼ inch [6 mm.] thick. Place the strips in marinade — here, a mixture of soy sauce, brown sugar, pepper, ground ginger and salt. Cover the strips and put them in the refrigerator overnight.

2 **Laying out strips.** Remove an oven shelf and cover it with a sheet of nylon netting — available at fabric shops — or cheesecloth to keep the meat from falling between the rungs. Remove the meat strips from the marinade, dry them on paper towels, and put them in one layer on the shelf, spacing them well apart.

3 **Packing the jerky.** To catch drips, place a foil-covered shelf in the lowest oven position; heat the oven to 140° F. [60° C.]. Put the meat on its shelf into the oven, leaving the door ajar and forcing circulation with a fan. If you use several shelves of meat, change their positions every two hours. Dry for eight hours, until brittle. Store in airtight containers.

Extracting Moisture with Dry Salt

Layered with salt, thin pieces of food render their juices, which dissolve the salt into a brine that prevents the growth of spoilage organisms. This method, known as dry-salting, produces various effects, depending on the foods involved.

Tiny fish may be dry-salted, for instance, to concentrate their flavors (top demonstration; recipe, page 162); the preserved fish become piquant seasonings for salads and meat dishes. Anchovies—small herring available fresh in spring and summer—are the favorites for such a treatment; other fish smaller than 5 inches [13 cm.] also can be used. The cleaned fish are given a preliminary salting to draw out excess juice, then are packed with salt and weighted so that they exude enough liquid to form brine. Oil will rise to the top of the packed vessels; the oil must be removed to prevent rancidity, after which the fish may be refrigerated for as long as two years.

A more complex reaction occurs when vegetables are dry-salted. While the resulting brine prevents spoilage, certain benign microorganisms on the vegetables continue to thrive. By fermentation, they convert vegetable sugars into acids, which add a delicious astringent flavor. If cabbage is used (bottom demonstration; recipe, page 156), the result is sauerkraut—"sour cabbage" in German.

For success, recipes must be followed precisely. If too little salt is used, the vegetables spoil; too much salt prevents fermentation. The fermenting cabbage must also be weighted to keep it under its brine. And the temperature of the area where you keep the fermenting vegetable must be controlled: Like salted fish, vegetables are best cured at temperatures below 70° F. [20° C.]; unlike fish, vegetables cannot be salted under refrigeration—temperatures below 65° F. [18° C.] will slow or stop fermentation.

During fermentation, a froth of carbon dioxide bubbles will appear on the sauerkraut. This must be removed regularly or it will interfere with fermentation. When fermentation is complete and bubbling has stopped—in about three weeks—the sauerkraut can be refrigerated for six to nine months; for longer storage, it should be processed (pages 30-31).

A Double Application for Anchovies

1 **Preparing the fish.** Line a dish with a layer of pure, coarse salt. Slit each fish belly open with a fingernail, then pinch off the head and pull it away with the guts. Rinse the fish and place them in the dish in layers, coating each layer with salt. Let the fish rest at room temperature for three to four hours.

2 **Drying the fish.** Fold a large piece of paper towel in half and place it on a work surface. Arrange the fish in a single layer on the towel. Press another folded piece of towel gently on top of the fish to blot up any excess moisture.

Controlling Fermentation in Sauerkraut

1 **Salting cabbage.** Rinse and shred green cabbage, and place it in a large, nonreactive bowl. Sprinkle in pickling or kosher coarse salt; toss with your hands. Let the cabbage rest for about five minutes so that the salt can begin drawing out the juice.

2 **Packing.** Layer the cabbage with coarse salt and spices such as juniper berries and peppercorns in a large, sterilized crock (page 86), packing it with a pestle. Leave a 3-inch [8-cm.] headspace. Pour on the remaining juice to cover. Then cover with a cloth, a plate and a weight, and store.

3 Packing the fish. Sterilize canning jars*(page 10)*. Sprinkle ¼ inch [6 mm.] of salt into each jar and place a layer of anchovies on top. Continue layering in this manner, leaving a ½-inch [1-cm.] headspace at the top of each jar and finishing with a salt layer.

4 Skimming. Weight the anchovies by inserting small, water-filled jars into the packing jars on top of the salt. Leave in a cool place for five or six days, then remove the weight and skim the fish oil from the brine. Close the jars, but do not heat them: This would destroy the fish's texture. Refrigerate the jars.

5 Filleting anchovies. Before use, soak the anchovies in water for at least 15 minutes to desalt them. Spread open a fish, then grasp its tail and peel the backbone away from the flesh; place the fillets on a paper towel to dry.

3 Removing froth. After a week, remove the coverings*(above, left)*. Using a ladle or skimmer, skim froth from the top of the brine*(right)*. With a clean cloth, wipe any scum from the crock rim. Cover with a clean cloth and add the cleaned plate and weight.

4 Completing the process. Skim the sauerkraut every day for about three weeks, until no more froth appears, indicating that the sauerkraut is ready. Cover the crock and refrigerate. Or, for canning, place the sauerkraut in a large, nonreactive pan and bring to a boil, stirring occasionally.

5 Canning. Fill clean, hot jars with the hot sauerkraut, packing it down to remove air bubbles and leaving a headspace of ½ inch [1 cm.]. Process the jars in a boiling water bath *(page 33)*.

The Special Uses of Brine

A salt-water brine, if properly handled, preserves and enhances a variety of foods whose size, shape or delicacy makes them unsuitable for dry-salting. Among the most fragile of these foods is caviar *(recipe, page 163)*, prepared by giving tiny fish eggs the briefest immersion in brine. The brine quickly penetrates the translucent eggs, firming them and brightening their subtle flavor.

To make the caviar, you can use the eggs—the roe—of any of a variety of fish. Sturgeon is the best-known source, but these fish are available only to West Coast sportsmen. Among the sorts of roe you can buy in the market, the best is that of salmon. Lacking salmon roe, use the roe of cod, tuna, whitefish or herring.

All of these roe come in a membranous sac. To free the eggs from the sac and also separate them from each other without breakage, slit open the sac, then rub it, cut side down, over a screen whose mesh is slightly wider than the eggs' diameter *(Step 2, top demonstration)*. The best mesh for this purpose is found in nylon netting or screening, available from fabric or hardware stores. Once separated and brined, the caviar will keep in the refrigerator for up to two months.

Hard green olives—ripe olives are too soft for the treatment described here—are another delicacy that responds well to brining *(recipe, page 154)*. Sold at specialty markets in large cities, the olives are available from September to November. They require a two-stage preparation: First, they are cracked and soaked in 10 changes of fresh water to remove their bitter taste; then they are brined.

For the demonstration here, the brined olives were kept in the refrigerator to prevent fermentation, thereby ensuring that their flavor remained mild. If you wish the olives to have the tart taste provided by fermentation, follow the procedure used for dill pickles *(pages 86-87)*. Treated either way, the olives will keep in the refrigerator for up to three months.

Brief Immersion to Make Caviar

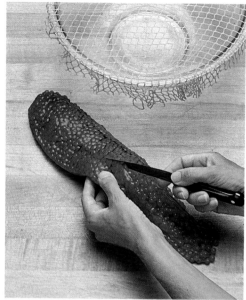

1 **Opening the roe sac.** Prepare a screen for separating the eggs — salmon roe, in this case: Tie ¼-inch [6-mm.] nylon mesh over a large shallow bowl to make a taut cover. Slide the tip of a small, sharp knife — its cutting edge up — into one end of the roe sac and draw the knife along the sac to slit it open.

2 **Separating the eggs.** Prepare a salt brine. Open the roe sac like a book along the slit and place it, open side down, on top of the nylon screen. Gently press and rub the sac across the screen; the eggs will fall into the bowl. If your fingers should stick to the roe sac, moisten them in the brine.

A Lengthy Steeping for Olives

1 **Tapping the olives.** Put the olives on a tray with sides that will prevent the fruit from skidding across the work surface. With a wooden mallet, lightly tap each olive to split its skin. Put the olives in a bowl of cold water. Cover the bowl and leave the olives in the refrigerator for 24 hours.

2 **Changing the water.** Drain the olives in a colander. Rinse the bowl, fill it with fresh water and tip in the olives. Cover the bowl and leave the olives in the refrigerator for 24 hours. Repeat the soaking-and-draining procedure every day for about nine more days.

3 **Brining the roe.** Add the separated eggs to the prepared brine. With your fingers, gently swirl the eggs to loosen any lingering bits of membrane. Let the eggs rest in the brine for 15 to 20 minutes; by the end of that time, they will have sunk to the bottom, leaving any sac residue floating on the surface. With your hand, carefully brush aside the residue and scoop up the eggs; transfer them to a strainer set over a bowl.

4 **Packing the eggs.** Let the eggs drain in the refrigerator for one hour to rid them of excess brine. Empty out the brine and tip the eggs into the bowl. Then spoon the eggs into small, sterilized jars, filling the jars as full as possible to eliminate air. Cover the jars with airtight lids, and store them in the refrigerator.

3 **Packing the olives.** Put the olives into a sterilized, nonreactive container, such as the glazed earthenware crock shown here. Fill the crock only about three quarters full in order to leave enough space for the brine.

4 **Adding brine.** Make enough brine to fill the container. If you like, add flavorings — in this instance, fresh bay leaves, dried fennel, coriander, thyme and orange peel. Bring the brine to a boil to enhance its flavor; cool it and strain it over the olives.

5 **Serving the olives.** Cover and refrigerate the container. The olives will be ready to eat after a week, but will keep for several months and slowly develop a more intense flavor. Use a perforated spoon to remove as many as you need at a time; then re-cover the crock and refrigerate it again.

Old-fashioned Dill Pickles Fermented with Flavorings

Like sauerkraut *(pages 82-83)*, traditional dill pickles acquire character—tangy taste and crisp texture—from the fermentation that occurs as they are preserved by salt. Because these pickles are made from vegetables too dense for dry salt to penetrate evenly and preserve effectively, they are immersed in a more penetrating salt-and-water brine, a process called long brining.

To contain the pickles and their brine, use a nonreactive vessel such as the glazed earthenware crock shown here. The glaze should be uncracked: Cracks allow leakage and may harbor harmful bacteria. Before you begin brining, prepare the crock by washing it thoroughly and rinsing it with scalding water.

Vegetables selected for long brining should be just-ripe and firm. Cucumbers, by far the favorite, should be the young pickling varieties used for short brining *(pages 62-63)*. It is important that any blossoms be removed: Blossoms contain enzymes that can soften the pickles. As extra insurance against softening, pack the cucumbers with fresh grape leaves, which contain a chemical compound that inhibits cellular breakdown.

The brine itself must be made with care. Salt preserves the pickles as they ferment; but too much or too little salt will stop fermentation, and the pickles will be soft and unpalatable. Most brines contain herbs and spices—dill is invariably present—and many include vinegar to retard the growth of unwanted microorganisms. Because too much vinegar will stop fermentation, follow recipes *(pages 128-129)* exactly.

Similar attention should be paid to storage temperature. If it is much below 65° F. [18° C.], fermentation will slow or stop. If it is much above 70° F. [20° C.], the pickles will overferment and bloat.

As with sauerkraut, the frothy scum that forms on the brine surface must be removed regularly. When frothing diminishes and the pickle flesh is firm and translucent—after three to six weeks—the pickles are ready. They can be stored for up to six months in the refrigerator. If the pickles are processed *(pages 30-31)*, they will keep indefinitely.

1 **Preparing cucumbers.** Rinse cucumbers in cold water. Scrub them with a brush to remove encrusted dirt, but scrub gently: Hard brushing could remove the bacteria that cause fermentation. Pinch off any stems from the ends of the cucumbers. Dry the cucumbers on paper towels.

5 **Removing scum.** In three to five days, scum will appear around the plate. Remove the crock coverings *(above, left)*, and ladle off the scum *(right)*. Wash the jar and plate, wrap the plate in fresh cheesecloth, and re-cover the crock. Cleanse daily until only a little scum appears, indicating slowed fermentation.

2 **Packing.** Pack the cucumbers into a clean crock, laying them on their sides. When the crock is half-full, add flavorings — here, bay leaves, dried chilies, dill seeds, sliced fresh ginger, whole cloves, mustard seeds, cinnamon sticks and peeled garlic cloves. Add fresh grape leaves, if available.

3 **Adding brine.** Pack the remaining cucumbers and flavorings into the crock. Bring salt and water to a boil; pour this brine into a bowl and let it cool. Ladle the brine over the cucumbers until they are completely covered.

4 **Weighting the pickles.** Choose a clean plate that will just fit into the crock and wrap the plate in several layers of cheesecloth. Fit the plate into the crock on top of the cucumbers. To keep the cucumbers completely submerged, weight the plate — a quart [1-liter] jar of water is used here. Set the crock aside.

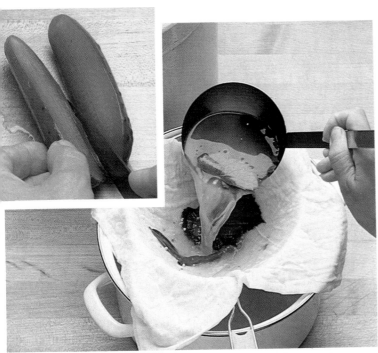

6 **Testing for doneness.** After three weeks, when foaming has reduced, slice open a pickle. If it is olive green throughout *(inset)*, the pickles are ready. If it shows white spots, let the pickles ferment for several more days, then test again. When they are ready, remove the pickles from the crock, and strain the brine through cheesecloth into a pan *(above)*.

7 **Canning.** Bring the brine to a boil; it will be cloudy, but richly flavored. Pack the pickles with fresh herbs and spices into clean, hot canning jars. Pour in the hot brine to within ½ inch [1 cm.] of the jars' rims. Process the pickles by putting the jars in a boiling water bath: Starting the timing immediately, process for 15 minutes.

Corned Beef: A Classic of the Preserver's Art

Cured in brine, meat will keep for about a month in the refrigerator. Since this storage period is relatively short, preservation is not the primary motive in the brining of meat: The technique is used principally to give the meat a distinctive flavor and texture.

Meats that are well flavored—beef, lamb, venison or pork, for instance—all react well to brining, and in Scandinavian countries, goose and chicken are also treated this way. While small pieces of meat could be effectively cured by dry-salting, large or irregularly shaped cuts can only be brined: The liquid will surround the meat, penetrating it quickly and evenly. Brining also is a good way to cure several pieces of meat together. But to avoid confusion of flavors, do not put meat from more than one kind of animal into the same brine.

Before you brine any meat, it is worthwhile dry-salting it briefly *(Step 3)*. This draws out excess blood that could dilute the brine, retarding the curing process.

The brine for meat can simply be salt and water. For added flavor, however, many cooks replace some or all of the water with wine or beer. They may also add sugar to impart a hint of sweetness, and herbs and spices for extra character *(recipe, page 167)*.

Brining time depends on the meat's thickness and on the intensity of flavor you want. A minimum of nine days is necessary to cure a piece of meat 1 inch [2½ cm.] thick. Longer brining will flavor the meat more assertively; if the meat, however, is left in the brine for more than a month, it will require overnight soaking to attenuate its saltiness before it is cooked. To contain the meat and brine, use a sterilized, nonreactive receptacle such as a glazed earthenware crock *(pages 82-83)*. During the brining, keep the crock in the refrigerator; a temperature no higher than 38° F. [3° C.] acts as insurance against spoilage.

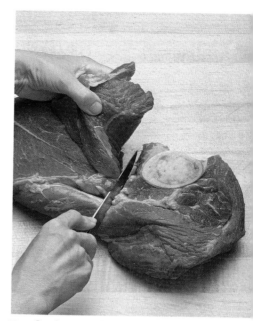

1 Cutting out the bone. Select whatever large cut and kind of meat you like—here, an arm roast of beef is used. With a sharp boning knife, cut close around the bone to separate each section of muscle. Trim and discard fat or gristle. Refrigerate or freeze the bone for soup or stock.

4 Making the brine. Start to make the brine about five hours before the dry-salting of the meat is finished. Put water and pure salt in a large stockpot. If you like, add other flavorings—in this case, brown sugar, crushed, unpeeled garlic cloves, dried thyme, bay leaves, and crushed black peppercorns and juniper berries. Bring to a boil, remove from the heat, and let the brine cool for four or five hours, so that the flavorings permeate the liquid.

5 Rinsing the dry-salted meat. Fill a large bowl with fresh, cool water. One by one, lift each slab of meat from the salt. Brush excess salt from the meat, then rinse the meat in the fresh water to clean off any caked salt and blood. The dry-salted meat will have darkened in color, shrunk slightly and become less pliable. Put the pieces of meat into a clean crock.

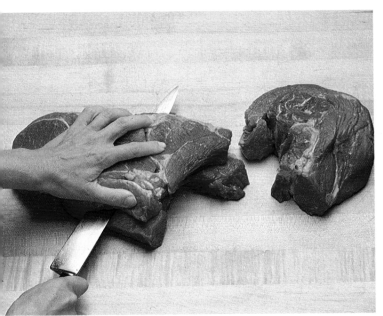

2 **Slicing the meat.** To keep curing time to a minimum, cut the meat into slabs about 1 inch [2½ cm.] thick. Holding the meat firmly against the work surface with the palm of one hand, use a large, sharp chef's knife to slice the meat horizontally.

3 **Dry-salting the meat.** Prick both sides of each slab in several places with a metal skewer or cake tester *(above)*. Sprinkle a ½-inch [1-cm.] layer of pure salt over the bottom of a large mixing bowl. Rub each slab of meat on both sides with salt; lay the pieces in the bowl one at a time, coating each piece of meat with a ½-inch layer of salt. Refrigerate for 24 hours.

6 **Straining in the brine.** Line a strainer with cheesecloth and rest it on top of the brining crock. Pour the cooled brine through the strainer to fill the crock to within 2 inches [5 cm.] of its rim. To keep the meat submerged, place a clean plate inside the crock; on the plate, put a large jar filled with water to serve as a weight.

7 **Testing for doneness.** Refrigerate the crock for nine or 10 days. To test whether the meat is completely cured, remove the weights and use tongs to lift out the topmost piece: It should be uniformly brown in color. The meat may be left in the brine for as long as several months, gaining a progressively stronger flavor.

Adding the Distinctive Tang of Smoke

Smoking preserves meat, poultry or fish just as drying does: Warm, dry air, in this case generated by smoldering wood, circulates around the food and gradually depletes it of moisture. In its results, however, smoking is a world apart from air-drying. The aroma of the wood gently scents the food, which, when first brined and salted to initiate drying and add flavor, is an incomparable product.

There are two kinds of smoking. Hot-smoking, in which food is hung close to the fire and cooked by its heat, does not preserve food. For that, you must cold-smoke it, hanging the food far enough from the fire that it is constantly surrounded by air in a temperature range of 70° to 90° F. [20° to 32° C.]—ideally, 85° F. [30° C.]. The easiest way to do this is to adapt a commercial hot-smoker, as demonstrated here. Hang the food above the chamber rather than inside it, and cover the food with a large box.

The wood you use will add flavor to the food—here, a trout. Mild-scented wood such as apple or alder gives a delicate flavor to fish; oak, hickory and beech suit more robust meats. Do not use resinous woods such as pine, which will produce an objectionable flavor. Smoking times vary from a few hours to a week or more, depending on the length of preliminary salting and brining, the size of the food, and its fat content. Smoking is complete when the food is firm and very dry.

Smoked meat or fish should be refrigerated loosely covered; airtight wrappings trap moisture and promote spoilage. Eat the food within six weeks: Serve it cold, or soak it in water for an hour to rehydrate it, then broil or bake it.

1 Dry-salting the fish. Gut the fish and rinse them well. Soak them for 30 minutes in brine. Rinse and drain thoroughly on paper towels. In a large bowl, layer the fish with a dry-salting mixture of equal quantities of salt and brown sugar. Cover and refrigerate the fish for two days.

2 Cleaning off the dry salt. Rinse the fish well, inside and out, in a bowl of fresh water or under cold running water. Put the fish on a rack over a jelly-roll pan. Refrigerate them for about four hours, until a glossy protein membrane called the pellicle forms on the skin; the membrane helps spread the smoke evenly over the fish.

3 Smoking the fish. Following the manufacturer's instructions, set up your smoker outdoors. About 15 minutes ahead, put a pan of wood chips — here, alder — in the smoker (left). Hang the fish on S hooks from the bar over the smoker, well above the heat (above). Cover the fish with a cardboard box to catch the smoke.

4 Testing for doneness. Let the fish smoke for about five days, replacing the wood chips periodically so that smoke is produced steadily. To determine whether the fish are fully cured, pry open the belly of one fish and gently squeeze the flesh (above): It should be dry — not juicy — and firm. Loosely wrap the fish in wax paper and place in the refrigerator.

Anthology
of Recipes

Drawing upon the cooking literature of 23 countries, the editors and consultants for this volume have selected 250 published recipes for the Anthology that follows. Many of the recipes were written by world-renowned exponents of the culinary art, but the Anthology also includes selections from rare and out-of-print books and from works that have never been published in English. Whatever the sources, the emphasis is always on techniques that are practical for the home cook.

Since many early recipe writers did not specify amounts of ingredients, types of pans or even cooking times and temperatures, the missing information has been judiciously added. Where preserves require processing in a boiling water bath or pressure canner, this fact has been indicated. In some cases, clarifying introductory notes have also been supplied; they are printed in italics. Modern recipe terms have been substituted for archaic language; but to preserve the character of the original recipes and to create a true anthology, the authors' texts have been changed as little as possible.

Apart from the primary components—fruits, vegetables, meat and fish—all ingredients are listed in order of use, with both the customary United States measurements and the metric measurements provided. All quantities reflect the American practice of measuring such solid ingredients as sugar by volume rather than by weight, as is done in Europe.

To make the quantities simpler to measure, many of the figures have been rounded off to correspond to the gradations on U.S. metric spoons and cups. (One cup, for example, equals precisely 240 milliliters; however, wherever practicable in these recipes, a cup's equivalent appears as a more readily measured 250 milliliters—¼ liter.) Similarly, the weight, temperature and linear metric equivalents have been rounded off slightly. Thus the American and metric figures do not exactly match, but using one set or the other will produce the same good results. However, because the moisture content of fruits and vegetables varies widely from one season or variety to another, many recipes may yield more—or less—than the quantities shown.

Unlike other recipes, those for preserving with sugar, salt, vinegar and alcohol demand an almost laboratory-perfect balance of components. Never alter the amounts of individual ingredients in such a recipe. However, these recipes can be halved, doubled and so on by decreasing or increasing the amounts of all of the ingredients proportionately. The adjusted amounts can be cooked the same way as the original amounts—except for jellies, marmalades, jams and the like: Do not cook more than 6 cups [1½ liters] of the combined sugar and fruit or fruit juice at once; larger batches may darken or lose flavor, or both.

Juices, Nectars and Syrups

Tomato Cocktail

To make about 2 quarts [2 liters]

8 lb.	ripe plum tomatoes, quartered (about 4 quarts [4 liters])	3½ kg.
1	medium-sized red or green sweet pepper, halved, seeded, deribbed, and cut into pieces	1
1	small onion, chopped	1
1	garlic clove, thinly sliced	1
1 or 2	celery ribs with leaves, chopped	1 or 2
½ tsp.	black peppercorns	2 ml.
2	small dried hot chilies, each about 1½ inches [4 cm.] long, or a 1-inch [2½-cm.] slice fresh hot chili (optional)	2
1 tsp.	coriander seeds or ½ tsp. [2 ml.] ground coriander	5 ml.
½	bay leaf	½
4 or 5	parsley sprigs	4 or 5
6 or 7	fresh basil leaves or ½ tsp. [2 ml.] crumbled dried basil	6 or 7
½ tsp.	mustard seeds	2 ml.
1 tsp.	sugar (optional)	5 ml.
1 to 2 tsp.	salt	5 to 10 ml.
	strained fresh lemon or lime juice	

Place the tomatoes in an enameled, tinned or stainless-steel pan. Add the sweet pepper, onion, garlic, celery, peppercorns, hot chili (if used), coriander, bay leaf, parsley, basil and mustard seeds. Bring to a boil, then simmer, stirring occasionally, until the vegetables are soft, about 15 minutes.

Force the vegetables through the finest disk of a food mill. Strain the juice if it is not smooth enough or if any seeds have passed into it. If the juice seems too thin, let it settle, then skim off and discard the thin liquid on top. Taste the juice for seasoning, and add sugar, salt, and strained lemon or lime juice to taste.

Return the juice to the rinsed-out kettle and bring it to a boil, then ladle it into clean, hot canning jars, leaving ½ inch [1 cm.] of headspace. Wipe the rims, put on two-piece lids, and fasten the screw bands.

Set the jars on a rack in a deep kettle half-filled with boiling water. Add boiling water to cover the jars by at least 2 inches [5 cm.]. Bring to a boil, cover, and process for 15 minutes. Remove the jars from the boiling water bath and allow them to cool.

HELEN WITTY AND ELIZABETH SCHNEIDER COLCHIE
BETTER THAN STORE-BOUGHT

Vegetable Juice Cocktail

To make about 9 quarts [9 liters]

6 to 8	large onions, chopped	6 to 8
4	large green peppers, halved, seeded, deribbed and cut into pieces	4
1	large bunch celery, sliced, the leaves reserved	1
¼ cup	salt	50 ml.
25 lb.	ripe tomatoes (½ bushel), quartered	12 kg.
2	bay leaves	2
12	whole cloves	12

Combine the onions, peppers and celery, and add the salt. Let the mixture stand for several hours or overnight.

To the tomatoes add the celery leaves, bay leaves and cloves. Combine the tomato mixture with the onion-pepper mixture. Heat until the juice from the tomatoes flows freely. Quickly put the vegetables into a food mill. Let all of the juice drain out. Then press out about half of the pulp.

Heat the juice and extracted pulp just to boiling. Pour into clean, hot jars. Leave ½ inch [1 cm.] of headspace. Cover. Process in a boiling water bath for 15 minutes.

BERNARDIN HOME CANNING GUIDE

Apricot Nectar

To make about 6 pints [3 liters]

7 lb.	ripe, perfect apricots, halved and pitted	3 kg.
1½ cups	water	375 ml.
3 cups	sugar	¾ liter
6 tbsp.	strained fresh lemon juice	90 ml.

In a saucepan, boil the water and sugar for two minutes, then let the syrup cool.

Purée the apricots through a sieve or food mill. You should have about 10 cups [2½ liters] of purée. Stir in the lemon juice and cool sugar syrup. Pour the nectar immediately into jars, filling them to within ½ inch [1 cm.] of the tops. These operations must be carried out as rapidly as possible in order to avoid any contamination of the nectar. Process for 15 minutes in a boiling water bath. Store the jars in a cool, dark place.

ANGELO SORZIO
THE ART OF HOME CANNING

Quince or Apple Nectar

Nektar ot Dyuli i Nektar ot Yabulki

To make about 3 pints [1½ liters]

2 lb.	quinces, coarsely grated (about 5 cups [1¼ liters]), or apples, cored and sliced (about 2 quarts [2 liters])	1 kg.
5 cups	water	1¼ liters
1 cup	sugar	¼ liter

Place the fruit in a large saucepan with 1 cup [¼ liter] of the water. Cover with a tight-fitting lid and simmer for 10 minutes, or until the fruit is quite soft. Press the juice and pulp through a fine sieve or through a food mill.

Heat the sugar and the remaining water in a saucepan over low heat, stirring until the sugar dissolves; then boil rapidly for a minute or two.

Measure the fruit purée, and add an equal volume of the sugar syrup. Stir thoroughly, cool, and process a small batch at a time in a blender. In a saucepan, bring the nectar mixture rapidly to a boil. Pour at once into jars or bottles and cover tightly. Process *(pages 30-31)*.

T. HADZHIYSKI, D. DONKOV, N. PEKACHEV, P. KOEN AND M. TZOLOVA
DOMASHNO KONSERVIRANE

Raspberry Syrup

Sirop de Framboise

To make 2 pints [1 liter]

1 lb.	raspberries (about 2 pints [1 liter])	½ kg.
about 4 cups	sugar	about 1 liter

In a bowl, crush the raspberries with a wooden pestle, then place the bowl in a cool place for 24 to 36 hours to allow the fruit to ferment.

Extract all of the juice from the raspberries by twisting them in a cloth. Measure the juice, and add 1 cup [¼ liter] of sugar to each ½ cup [125 ml.] of juice. Place over low heat and bring to a boil, stirring frequently. Remove from the heat as soon as the mixture comes to a boil. Bottle the mixture and process *(pages 30-31)*.

MYRETTE TIANO
LES CONSERVES

Black Currant Liqueur

Liqueur de Cassis

To make about 2 quarts [2 liters]

2 lb.	black currants, with a few leaves (about 7 cups [1¾ liters])	1 kg.
4 cups	brandy or *eau de vie*	1 liter
1½ cups	sugar	375 ml.
2 cups	water	½ liter

Put the bunches of black currants, with their leaves, in a widemouthed glass jar or a stoneware bottle. Pour in the alcohol, cover tightly and let macerate for about two months.

Drain the flavored alcohol from the fruit. Mix the sugar and water, and bring just to a boil. Cool this syrup and add it to the alcohol. Bottle and cork. Keep the bottles of liqueur for one month before opening.

GINETTE MATHIOT
JE SAIS FAIRE LES CONSERVES

Essence of Vanilla

This essence is a richly flavored version of vanilla extract. To substitute the essence for extract, pour it from the bottle through a tea strainer and measure out half of the usual quantity of flavoring.

To make 1 pint [½ liter]

3	vanilla beans, cut into very small shreds	3
2 cups	brandy	½ liter

Put the vanilla shreds into a bottle with the brandy, and cork the bottle. Shake the bottle occasionally, and in the course of three months' time the essence will be ready for use.

OSCAR TSCHIRKY
THE COOK BOOK BY "OSCAR" OF THE WALDORF

Plain Grape Juice

To make about 4 pints [2 liters]

7 lb.	ripe, perfect white grapes	3½ kg.
3 cups	sugar	¾ liter

Press grapes through a sieve or food mill to extract their juice; discard the seeds and skins. Pour the juice into an enameled, tinned or stainless-steel pan. Boil the juice for five minutes. Stir in the sugar. Strain the juice mixture through a jelly bag, or a colander lined with a linen cloth or a double thickness of cheesecloth. Pour the juice into jars or bottles. Cover and process *(pages 30-31)*.

ANGELO SORZIO
THE ART OF HOME CANNING

Strawberry, Raspberry or Sour-Cherry Nectar

Nektar ot Yagodi, Malini, Vishni

To make about 4 pints [2 liters]

2 lb.	strawberries or raspberries, hulled, or sour cherries, pitted (about 6 to 8 cups [1½ to 2 liters])	1 kg.
2 cups	sugar	½ liter
4 cups	water	1 liter

Force the fruit through a fine sieve into a bowl. Cover the bowl. Prepare the syrup by dissolving the sugar in the water over low heat, then boil briskly for a minute or two. Measure the fruit pulp and add an equal volume of the sugar syrup. Stir together in an enameled, tinned or stainless-steel pan to blend, and bring rapidly to a boil over high heat. Fill hot bottles to within ½ inch [1 cm.] of the top with the bubbling nectar and cover immediately. Process *(pages 30-31)*.

T. HADZHIYSKI, D. DONKOV, N. PEKACHEV, P. KOEN AND M. TZOLOVA
DOMASHNO KONSERVIRANE

Jellies and Marmalades

Apple, Crab Apple or Quince Jelly

To make about 5 cups [1 ¼ liters]

4 lb.	tart apples, crab apples or quinces, quartered, stems and blossom ends removed	2 kg.
	sugar	

Place the fruit in a saucepan. Add water until it can be seen through the top layer of fruit. Cook, uncovered, until the fruit is soft, about 40 minutes. Wet and wring out a jelly bag and pour in the fruit and juice. Let drip through the bag without squeezing.

Measure the juice and put it into a large enameled, tinned or stainless-steel pan. Simmer the juice, uncovered, for about five minutes, skimming off any froth that forms. Add ¾ to 1 cup [175 ml. to ¼ liter] of sugar to each cup [¼ liter] of juice, and stir until the sugar is dissolved. Boil the mixture until the jelling point is reached. Pour the jelly into hot, dry jars. Seal.

IRMA S. ROMBAUER AND MARION ROMBAUER BECKER
JOY OF COOKING

Crab Apple Jelly

To make about 14 cups [3 ½ liters]

8 lb.	large crab apples, quartered (about 7 quarts [7 liters])	3½ kg.
¾ cup	strained fresh lemon juice	175 ml.
	sugar	

Cover the apples to a depth of 1 to 2 inches [2 to 5 cm.] with water, and cook to a mush, about 45 minutes. Pour into a coarse cotton bag or strainer, and when cool enough, press or squeeze hard to extract all the juice. Take a piece of fine muslin, spread over a colander placed over a crock, and with a cup, pour the juice slowly in, allowing plenty of time for the juice to run through; repeat this process twice, rinsing out the muslin frequently.

Measure the juice. Measure 2 cups [½ liter] of sugar for each 2½ cups [625 ml.] of juice. Add the lemon juice to the apple juice and boil for 10 to 20 minutes; while boiling, sift in the sugar slowly, stirring constantly. Boil for five minutes longer. This is generally sufficient, but it is always safer to try it and ascertain whether it will jell. Put in jars and cover. This makes a very clear, sparkling jelly.

THE BUCKEYE COOKBOOK

Apple or Quince Jelly

Apfel- oder Quittengelee

To make about 14 cups [3 ½ liters]

9 lb.	unripe apples or ripe quinces, peel and core included, sliced	4 kg.
	sugar	
3 tbsp.	strained fresh lemon juice	45 ml.
1 tsp.	vanilla extract or ⅔ cup [150 ml.] white wine	5 ml.

Cover the apples or quinces with water, cook for about 30 minutes, or until soft, and let stand overnight. The next day, pour the fruit into a jelly bag, and let the juice run through without stirring or pressing.

Measure the juice, and add ¾ cup [175 ml.] of sugar for each cup [¼ liter] of juice. Cook until the jelling point is reached (the jelly will turn amber-colored), skim thoroughly, and stir in the lemon juice and vanilla extract or wine.

Pour the jelly into jars while still hot and seal.

ELIZABETH SCHULER
MEIN KOCHBUCH

Currant Jellies

Gelée de Groseilles

To make about 28 cups [7 liters]

13 lb.	red currants, or a mixture of red and white or black currants	6 kg.
2 lb.	raspberries (about 4 pints [2 liters])	1 kg.
30 cups	sugar (15 lb. [6¾ kg.])	7½ liters

In a large bowl, arrange the currants in layers, sprinkling each layer heavily with sugar. In another bowl, layer the raspberries with the remaining sugar. Let the fruits macerate for five to six hours.

In an enameled, tinned or stainless-steel pan, cook the sugared currants over very low heat, stirring with a wooden spatula until the mixture reaches a boil. With a spoon, scoop up a few drops of the juice and drop them into cold water; if the juice jells, the currants are cooked. Add the raspberries and return the currants to a boil for a minute or two.

Pour the mixture into a large, fine sieve or jelly bag set over a bowl, and without pressing on the fruits, let the mixture drip through. When nothing more drips from the sieve or bag, put the strained jelly into jars and let it cool. For a second batch of jelly, press down on the fruits remaining in the sieve, or squeeze the bag hard until you can extract no more juice. Place this smaller batch of jelly in jars. This jelly will be less transparent and delicate than the first batch because the squeezing will have extracted some bitterness from the currant seeds.

JULES BRETEUIL
LE CUISINIER EUROPÉEN

Currant Jelly

To further flavor this jelly, add some elder flowers tied in cheesecloth and boil them with it. For each 2½ cups [625 ml.] of juice, add a bunch of fresh elder flowers carefully picked from their stalks; if the bunches are large, less will do.

To make about 10 cups [2 ½ liters]

6 to 7 lb.	ripe white or red currants, freshly picked (about 5 quarts [5 liters])	3 kg.
about 8 cups	sugar	about 2 liters
about 2 cups	warm water	about ½ liter

Pick all the best ripe currants on each stalk and bruise them well on a plate. As they are done, put them in a cloth bag to drip; leave them so all night.

Measure the juice, and add 2 cups [½ liter] of sugar for each 2½ cups [625 ml.] of juice. Allow ½ cup [125 ml.] of water for each 2 cups [½ liter] of sugar.

Put the sugar and water into an enameled, tinned or stainless-steel pan over high heat and, when the sugar is dissolved, boil hard for about 10 minutes, or until the temperature is 250° F. [120° C.]. Add the currant juice and boil for three to four minutes. Pour the jelly into jars and cover.

CATHERINE FRANCES FRERE (EDITOR)
THE COOKERY BOOK OF LADY CLARK OF TILLYPRONIE

Red Currant Jelly

Gelée de Groseilles

To make about 8 cups [2 liters]

6 lb.	red currants (about 5 quarts [5 liters])	3 kg.
½ cup	water	125 ml.
about 4 cups	sugar	about 1 liter

Put the currants and the water in an enameled, tinned or stainless-steel pan. Cook over medium heat for about eight minutes, stirring constantly and breaking the currants against the sides of the pan. Pour the contents of the pan into a fine sieve or jelly bag, and let the juice drip into a bowl. Do not press the fruit pulp if you want a perfectly clear jelly.

Measure the juice. Put an equal amount of sugar into the preserving pan, and add the juice. Place over low heat, and stir constantly until the mixture reaches the boiling point. As soon as it boils, stop stirring. Skim the mixture, and let it continue to boil for exactly three minutes—the time necessary for the currant pectin to combine with the sugar and set to a jelly. Put the jelly into jars and cover while hot.

GINETTE MATHIOT
JE SAIS FAIRE LES CONSERVES

Elderberry and Apple Jam

To make about 2 ½ pints [1 ¼ liters]

1 lb.	elderberries (about 3 ½ cups [875 ml.])	½ kg.
1 lb.	apples, coarsely chopped, including the peel, core and seeds (about 4 cups [1 liter])	½ kg.
1 ¼ cups	water	300 ml.
4 to 5 cups	sugar	1 to 1 ¼ liters

Put the fruits and water into a large ovenproof dish or pan, cover, and bake in a 275° F. [140° C.] oven until the juice runs from the fruit and the fruit becomes pale and soft, about two hours. Then strain off the juice, pressing lightly to extract the juice without pulping the fruit.

Measure the juice; for each 2½ cups [625 ml.] of juice, allow 2 cups [½ liter] of sugar. Put the juice and sugar into an enameled, tinned or stainless-steel pan, and boil rapidly for about 30 minutes, or until the jelling point is reached. Put into jars and cover. Process *(pages 30-31)*.

MARY AYLETT
COUNTRY FARE

Stanhope Black Jelly

This is delicious if you take care to do it right and you get none of the small seeds (elderberries are full of them).

To make about 8 cups [2 liters]

3 ½ lb.	ripe elderberries (about 3 quarts [3 liters])	1 ¾ kg.
1 ½ lb.	crab apples or small green apples, halved (about 1 ½ quarts [1 ½ liters])	¾ kg.
5 cups	water	1 ¼ liters
	sugar	
6 tbsp.	strained fresh lemon juice	90 ml.
1 tbsp.	butter	15 ml.

Boil the elderberries in half of the water for 10 minutes. Boil the apples in the remaining water for 15 minutes. Combine the two mixtures and boil together for five minutes. Strain through a jelly bag.

Measure the juice. Allow 1 cup [¼ liter] of sugar for each 1¼ cups [300 ml.] of juice. Put the sugar in a preheated 300° F. [150° C.] oven for about five minutes to get hot.

To the elderberry and apple juice, add the lemon juice and butter. Bring to a full rolling boil, then add the hot sugar. Stir thoroughly and boil fast. Skim and test after five minutes and, when it jells, put into jars and seal.

PEGGY HUTCHINSON
PEGGY HUTCHINSON'S PRESERVING SECRETS

Mint, Sage and Gooseberry Jelly

To make about 5 cups [1 ¼ liters]

2 lb.	gooseberries, halved (about 6 cups [1 ½ liters])	1 kg.
6	sprigs mint, 4 leaves very finely chopped	6
2 tsp.	finely chopped fresh sage leaves	10 ml.
4 cups	water	1 liter
3 tbsp.	strained fresh lemon juice	45 ml.
	sugar	

Put the gooseberries in an enameled, tinned or stainless-steel pan with the water and lemon juice, bring to a boil, reduce the heat and simmer gently for 20 minutes. Put the mixture into a sieve lined with muslin or cheesecloth, and let drip through overnight.

Measure the gooseberry liquid, and combine with 2 cups [½ liter] of sugar for each 2½ cups [625 ml.] of liquid in a large, heavy pan. Bring to a boil, and boil over high heat for five minutes, or until a small spoonful of the jelly dropped on a saucer will set. Add the mint sprigs to the jelly as soon as it comes off the heat, and let infuse for 10 minutes. Strain the jelly through a fine sieve.

Steep the chopped mint and sage in enough boiling water to cover for five minutes, drain, and add the herbs to the jelly. Pour into jars, cover and store in a cool, dark place.

MARIKA HANBURY TENISON
RECIPES FROM A COUNTRY KITCHEN

Grape Jelly

To make about 6 cups [1 ½ liters]

4 lb.	ripe grapes	2 kg.
½ cup	water	125 ml.
about 4 cups	sugar	about 1 liter

Wash, stem, and crush the grapes. Add a small amount of water, bring to a boil and simmer for 15 minutes. Strain through a jelly bag into a ceramic or glass bowl, and let the juice stand overnight in a cool place to allow any formation of tartaric acid crystals to settle to the sides and bottom of the bowl. In the morning, pour off the juice carefully, leaving the sediment in the bottom of the bowl. Measure the juice and bring it to a boil. For each cup [¼ liter] of juice, add ¾ cup [175 ml.] of sugar. Boil the mixture rapidly to the jelling stage. Pour into sterilized glasses and seal.

ANN SERANNE
THE COMPLETE BOOK OF HOME PRESERVING

Wine Jelly

To make about 12 cups [3 liters]

3 lb.	ripe white grapes, crushed (about 2½ quarts [2½ liters])	1½ kg.
1¼ cups	dry white wine	300 ml.
1½ lb.	tart apples, sliced (about 6 cups [1½ liters])	¾ kg.
1	lemon, thinly sliced	1
6	cardamom seeds	6
	sugar	
¾ cup	brandy	175 ml.

Put the grapes and white wine in a large enameled, tinned or stainless-steel pan. Bring to a boil over high heat. Reduce the heat, and simmer for 20 to 30 minutes, until the grapes are soft and pulpy. Add the apples, lemon slices and cardamom seeds, and simmer for another 20 to 30 minutes, or until the apples are pulpy. Pour the contents of the pan into a jelly bag, and let the juice drip through for a minimum of 12 hours or overnight.

Measure the juice and pour it into a clean pan. Add 1 cup [¼ liter] of sugar for each 1¼ cups [300 ml.] of juice. Stirring constantly, cook over low heat until the sugar is dissolved. Increase the heat and boil briskly without stirring, until the jelling point is reached. Add the brandy, cook for another minute or two, and pour the jelly into jars. Cover.

MARYE CAMERON-SMITH
THE COMPLETE BOOK OF PRESERVING

Orange Jelly and Jam

Marmelade et Gelée d'Orange

To make about 6 cups [1½ liters] jelly and 5 pints [2½ liters] jam

11 lb.	oranges, peeled, sliced and seeded (about 7 quarts [7 liters])	5 kg.
	sugar	
2 to 2½ lb.	apples, cut up (about 2 quarts [2 liters])	1 kg.

Put the sliced oranges in an enameled, tinned or stainless-steel pan set in hot water over low heat for at least two hours, or until they are very soft and pulpy. Pour the oranges into a jelly bag and let their juice drip through without squeezing the bag. Reserve the remaining pulp. Measure the juice. Then measure an equal quantity of sugar, and cook the sugar with 2 tablespoons [30 ml.] of water over medium heat until it melts and forms a thick syrup. Add the orange juice and cook rapidly to the jelling point. Put the jelly into jars.

Cook the apples over medium heat until they are soft. Purée them through a sieve. Sieve the orange pulp that remains in the jelly bag, and for each 2 cups [½ liter] of orange purée, add 1 cup [¼ liter] of apple purée. Measure out enough sugar to equal the volume of the combined fruit purées. Stir the sugar into the purées and cook rapidly to the jelling point. Put the jam into jars. Process *(pages 30-31)*.

LA VARENNE
LE VRAY CUISINIER FRANÇOIS

Beach-Plum Jelly

The beach plum is a native American fruit that grows wild on low, sprawling bushes found along Atlantic beaches stretching from Maine to Delaware. The plums usually ripen in August and September.

To make about 3 cups [¾ liter]

2 lb.	ripe beach plums, stemmed (about 5 cups [1¼ liters])	1 kg.
	water	
	sugar	

Place the plums in an enameled, tinned or stainless-steel saucepan. Add enough water to cover them, bring to a boil and cook covered for five minutes. Then crush the plums with a potato masher and cook again until they are mushy, stirring from time to time. Place the mashed plums in a jelly bag and let them drip thoroughly. Boil the juice for five minutes, then measure it. Add an equal volume of sugar. Bring to a rapid boil, then simmer until the liquid will jell, about 40 minutes. Seal in jars while hot.

VERA GEWANTER AND DOROTHY PARKER
HOME PRESERVING MADE EASY

Damson Jelly with Basil

Serve this jelly with veal or poultry. It keeps well.

To make about 6 cups [1½ liters]

5 lb.	damson or sour plums (about 5 quarts [5 liters])	2½ kg.
3 cups	fresh basil leaves or 1½ cups [375 ml.] dried basil	¾ liter
6 cups	water	1½ liters
	sugar	

In a large enameled, tinned or stainless-steel pan, combine the plums, water and two thirds of the basil leaves. Boil together for about 30 minutes, or until the plums are pulpy and soft. Strain overnight through a jelly bag. For every 2½ cups [625 ml.] of juice obtained, add 2 cups [½ liter] of sugar. Stir over low heat, without boiling, until the sugar is dissolved, then boil until a jell is obtained. Skim. Chop and add the remaining basil, and allow the jelly to rest for about five minutes before putting into jars.

PATRICIA HOLDEN WHITE
FOOD AS PRESENTS

Orange-Lemon Jelly

To make 6 cups [1 ½ liters]

6	oranges, sliced	6
1 ¼ cups	strained fresh lemon juice	300 ml.
3 ¾ cups	water	925 ml.
	sugar	

Combine the lemon juice and water. Add the sliced oranges and let stand overnight. In the morning, bring the mixture to a boil, boil until the fruit is tender, and strain the liquid through a jelly bag. Measure the liquid and bring it to a boil. Add 1¾ cups [425 ml.] of sugar to each 2 cups [½ liter] of juice, and boil rapidly to the jelling stage. Pour into sterilized glasses and seal.

ANN SERANNE
THE COMPLETE BOOK OF HOME PRESERVING

Pomegranate Jelly

To make about 2 cups [½ liter]

12	underripe pomegranates	12
	sugar	

Place the pomegranates in a covered earthenware vessel. Bake in a preheated 200° F. [100° C.] oven for at least one hour, or until they become quite pulpy and the juice flows freely. Strain without any pressure through a jelly bag, which can be left to drip all night.

Measure the juice, boil it up, and add 1 cup [¼ liter] of sugar for each cup of juice. When the sugar has dissolved, boil up for a few minutes until the jelly will set. It is a lovely color. Pour into glass jars and cover.

MAY BYRON
MAY BYRON'S JAM BOOK

Quince Jelly and Paste

Gelée et Pâte de Coings

*To make about 8 cups [2 liters] jelly and
4 pounds [2 kg.] paste*

9 lb.	quinces, quartered, cores removed and tied in a cloth bag (about 5½ quarts [5½ liters])	4 kg.
	sugar	

Put the quince quarters and the bag of cores in an enameled, tinned or stainless-steel pan. Cover with water. Cook over medium heat for about 45 minutes, or until the quinces are easily pierced by a straw or skewer. Pour the contents of the pan into a strainer, and let all of the juice drip through into a clean pan. Measure the juice and add an equal amount of sugar. Boil for 10 minutes, or until the jelling point is reached. Skim and put the jelly into jars. Seal.

To make the quince paste, place the contents of the strainer in a clean pan and remove the bag of cores. Crush the quince pulp with a wooden pestle. Measure the crushed pulp, and add the same amount of sugar. Stir with a wooden spatula, then place over medium heat, and stir constantly until the paste thickens and pulls away from the sides of the pan, about 25 minutes.

Pour the paste onto deep plates to form rounds about 1½ inches [4 cm.] thick. Put the plates in a cool, well-ventilated place to dry for three to five days, turning the rounds over at least once during that time. Cut the paste into small pieces, roll them in sugar, and place them on wax paper to dry further. After several days, put them into airtight jars or tins, stacking them in layers with wax paper between; they will keep for months.

HENRI PHILIPPON
CUISINE DU QUERCY ET DU PERIGORD

Rowanberry Jelly

Rowanberries are the red fruit of the rowan, or mountain ash, tree. They ripen in early fall. Raspberries, blackberries, elderberries and blueberries are among the other berries that are also suitable.

To make about 10 cups [2 ½ liters]

2 lb.	rowanberries (or a mixture of rowanberries and other berries) (about 7 cups [1 ¾ liters])	1 kg.
2 lb.	crab apples or tart cooking apples, cut into chunks (about 8 cups [2 liters])	1 kg.
3 tbsp.	strained fresh lemon juice	45 ml.
	sugar	

Put all of the fruits in an enameled, tinned or stainless-steel pan, and add cold water just to the top level of the fruit. Add the lemon juice. Bring to a boil and simmer until the fruits are tender and mushy, about 40 minutes. Stir occasionally with a wooden spoon. Pour the contents of the pan into a jelly bag, and let the juices drip for six hours or overnight. Do not squeeze the bag; this makes the jelly cloudy.

Measure the juice, and allow 1 cup [¼ liter] of sugar for every 1¼ cups [300 ml.] of juice. Warm the sugar on the middle shelf of a 225° F. [110° C.] oven for 10 to 15 minutes. This step reduces the cooking time of the jelly, giving a fresher flavor. Heat the measured juice in the preserving pan and, when it reaches a simmer, add the sugar and stir until it dissolves. Boil rapidly until the jelling point is reached, in 10 to 12 minutes. Skim, put into jars and seal.

PETITS PROPOS CULINAIRES III

Strawberry and Orange- or Apple-Pectin Jelly

The technique of extracting juice from uncooked fresh berries is demonstrated on page 38. A recipe for orange pectin appears on page 101, one for apple pectin appears on page 165.

To make about 4 cups [1 liter]

2 cups	strawberry juice, extracted from fresh ripe berries	½ liter
2 cups	orange or apple pectin	½ liter
2 cups	sugar	½ liter

In an enameled, tinned or stainless-steel saucepan, mix the strawberry juice with the pectin. Add the sugar. Boil rapidly until the jelling point is reached. Skim and pour into hot jelly glasses. When cold, pour hot paraffin over the jelly.

LOUISIANA STATE UNIVERSITY COOPERATIVE EXTENSION
LOUISIANA COOPERATIVE EXTENSION PUBLICATION NO. 1568

Tomato and Red-Pepper Jelly

Tomaten-Paprika-Gelee

Tomato and red-pepper jelly is an accompaniment for cold dishes. It can be turned out of the jar and cut into decorative strips or cubes. It can be whipped up with a little red wine and served as a sauce for beef fondue, or stirred into fresh cream and used as a dressing for chicken salads.

To make about 6 cups [1 ½ liters]

1 ½ lb.	tomatoes, cut into large pieces	¾ kg.
1 lb.	long red sweet frying peppers, stemmed, quartered, seeded and coarsely chopped	½ kg.
2	large onions, coarsely chopped	2
6 tbsp.	strained fresh lemon juice	90 ml.
¼ cup	red wine	50 ml.
1 tsp.	salt	5 ml.
4 cups	sugar	1 liter

Purée the tomatoes and sweet peppers in the blender, or put through the fine disk of a food grinder. Put the purée into a pan with the onions and the lemon juice. Bring the mixture to a boil and boil for 10 minutes. Then pour into a jelly bag or a fine sieve, and let all of the juice drip into a large bowl.

Return the juice to a clean pan and add the wine, salt and sugar. Stirring constantly, bring to a boil, then let boil for two minutes, or until the jelling point is reached. Pour the jelly into jars and seal immediately.

EIKE LINNICH
DAS GROSSE EINMACHBUCH

A Fine Marmalade of Cherries

To obtain the raspberry and red currant juices for this recipe, place about 1 ½ quarts [1 ½ liters] of raspberries in one enameled, tinned or stainless-steel pan and about 1 ¾ quarts [1 ¾ liters] of red currants in another. Heat both the fruits gently for about 20 minutes, or until they are very soft and juicy. Then strain the fruits, separately, through jelly bags. The marmalade should be processed (pages 30-31).

Madame Mancy, who made this recipe in 1682 for Queen Mary of Modena, the wife of King James II, says, "Peradventure to keep all the year there may be requisite a little more sugar," but my experience is that it is eaten so greedily there is never any chance of keeping it a year. The quantity of sugar I have quoted gives it a pleasant tart flavor.

To make about 4 pints [2 liters]

3 lb.	cherries, pitted (about 3 quarts [3 liters])	1 ½ kg.
1 cup	raspberry juice	¼ liter
2 cups	red currant juice	½ liter
about 2 cups	sugar	about ½ liter

Put all the ingredients together into an enameled, tinned or stainless-steel pan; boil them over high heat, skimming off any scum that rises. When you find them of a fit consistency, with a fine clear jelly mingled with the cherries, take them from the heat, and bruise the cherries with the back of your spoon. Put them in jars.

MOIRA MEIGHN
THE MAGIC RING FOR THE NEEDY & GREEDY

Montserrat Marmalade

To help the marmalade jell, put the citrus seeds in a cheesecloth bag, and add them to the fruit when it is combined with the water. Remove the bag of seeds before adding the sugar.

To make 10 pints [5 liters]

1	grapefruit, unpeeled, sliced	1
1	orange, unpeeled, sliced	1
1	lime, unpeeled, sliced	1
1	pineapple, peeled, cored and cubed	1
2 quarts	water	2 liters
7 cups	sugar	1 ¾ liters

In an enameled, tinned or stainless-steel pan, combine the fruit with the water, cover and let the fruit soak overnight. The next day, uncover and boil for 15 minutes. Set aside for 24 hours. Add the sugar; boil until the mixture jells, about one hour. Pour into hot jars and process *(pages 30-31).*

CONNIE AND ARNOLD KROCHMAL
CARIBBEAN COOKING

Lemon Marmalade

When making this marmalade, add about half of the lemon juice and grated lemon peel to the clear syrup. Boil for five minutes, then taste it, and add more juice or peel if desired.

To make about 5 pints [2 ½ liters]

12	lemons, the peel grated, the juice strained and the shells soaked in cold water for 24 hours	12
6 cups	sugar	1 ½ liters
3 ½ cups	water	875 ml.

Boil the lemon shells in the soaking water until they are so tender that a straw can pierce them—about 45 minutes. Drain the shells and cut them into very thin strips about 1 inch [2½ cm.] long. Make a syrup with the sugar and water, and boil for about 10 minutes. Add the strips of lemon and boil until clear, about 15 minutes. Remove and reserve the lemon strips, and add as much of the lemon juice and grated peel as liked. When boiled thick, about 20 minutes, put in the lemon strips and boil for a further 10 minutes. Put into glass jars, cover and process *(pages 30-31)*.

MISS TYSON
THE QUEEN OF THE KITCHEN

Lemon, Watermelon and Bitter-Orange Jam

Confiture de Citrons, de Pastèques et d'Oranges Amères

Bitter oranges, also known as Seville or sour oranges, have a tart flavor. They are obtainable at some fruit specialty stores and Latin American markets in winter. The watermelon rind can be reserved for use in pickles.

This jam is not only delicious, but remarkable for its digestive powers. It looks pleasing, too: The melon pieces are suspended in a translucent orange syrup.

To make about 4 pints [2 liters]

3	lemons, the peels only, thinly pared and cut into julienne	3
one 4-lb.	watermelon, quartered, the flesh cut out, cubed and seeded (about 1 ½ quarts [1 ½ liters] of cubes)	one 2-kg.
4 or 5	bitter oranges, the peels of 3 of them thinly pared and cut into julienne, the juice of all 4 or 5 strained	4 or 5
¾ cup	water	175 ml.
	sugar	

Bring a large pan of water to a boil, add the lemon and orange peels, and boil for 10 to 15 minutes. Drain the peels.

Put the watermelon cubes in an enameled, tinned or stainless-steel pan with the ¾ cup [175 ml.] of water, and cook over low heat for about 20 minutes. Add the blanched peels and the orange juice. Measure the mixture, and add 3 cups [¾ liter] of sugar for each 4 cups [1 quart] of fruit.

Cook over very low heat (evaporation should be slow) until the cubes are transparent and the syrup has reached the jelling point, about one hour. Put in jars and cover immediately. Process *(pages 30-31)*.

ÉLIANE THIBAUT COMELADE
LA CUISINE CATALANE

Lemon and Apple Marmalade

To make about 7 pints [3 ½ liters]

3	lemons, thinly sliced and seeded	3
3 lb.	tart apples, peeled, cored and sliced (about 3 ¼ quarts [3 ¼ liters])	1 ½ kg.
7 ½ cups	cold water	1 ¾ liters
10 cups	sugar	2 ½ liters

Soak the lemons in the water overnight. Pour the lemons and water into an enameled, tinned or stainless-steel pan and cook gently for about 20 minutes, or until the peel is tender. Add the apples and cook until they are tender, approximately 20 minutes.

Stir in the sugar and boil until the jelling point is reached. Pour into jars and cover. Process *(pages 30-31)*.

GERTRUDE MANN
THE APPLE BOOK

Oxford Marmalade

A whole lemon, cooked and cut up with the oranges, may be used instead of the lemon juice, if desired. An alternative method of extracting the pectin from the seeds is to tie them in a muslin bag, add the bag to the marmalade at the point when the seed water would otherwise be strained in, then remove the bag when the jelling point is reached.

This famous chunky marmalade is a beautiful red gold, but can be made considerably darker, in the true approved Oxford fashion, by adding 1 ½ tablespoons [22 ml.] of black-strap molasses to the sugar syrup with the orange chunks.

To make about 5 pints [2 ½ liters]

12	small, fresh, unwrinkled bitter oranges	12
3 quarts	water	3 liters
8 cups	sugar	2 liters
3 tbsp.	strained fresh lemon juice	45 ml.

Put the oranges in a large pan with 2½ quarts [2½ liters] of the water, cover and let them boil for one hour. Take out the oranges and keep the water in which they were cooked.

Cut each orange in half and carefully scoop the seeds into a jug containing the remaining water.

Slice the oranges and pulp on a board, making the chips as thick and chunky as you like—about ½ by ¼ inch [1 by ½ cm.] is usual. Put the oranges' cooking liquid and the sugar in an enameled, tinned or stainless-steel pan, and heat gently, stirring all the time with a wooden spoon until the sugar has completely dissolved.

Now turn up the heat and boil for five minutes. Throw the orange chunks and lemon juice into the liquid. Strain the seeds' soaking water into the pan, pressing and rubbing the seeds in the sieve with the wooden spoon to extract their pectin, essential to the setting of jam and marmalade. Stir everything together and bring to a boil. This marmalade should jell in about 30 to 60 minutes. When the jelling point is reached, skim well, then pour the marmalade into warm jars and cover. Process *(pages 30-31)*.

CAROLINE CONRAN
BRITISH COOKING

Bitter-Orange Marmalade

Confiture d'Oranges Amères

Bitter oranges—also known as Seville or sour oranges— have a tart flavor. They are obtainable at some fruit specialty stores and Latin American markets November to January.

To make about 8 to 10 pints [4 to 5 liters]

12	bitter oranges, very thinly sliced, seeds removed and reserved	12
2	sweet oranges, very thinly sliced, seeds removed and reserved	2
1	lemon, very thinly sliced, seeds removed and reserved	1
3 tbsp.	fresh lemon juice	45 ml.
4½ quarts	water	4½ liters
	sugar	

Put the orange and lemon seeds into a small bowl of water; set aside. Put the sliced oranges and lemon into an enameled, tinned or stainless-steel pan with the water, and let steep for 24 hours.

Place the pan over medium heat, and boil the orange and lemon slices for about 35 minutes. Remove from the heat and let steep for another 24 hours.

Measure the fruits and their liquid, and return them to the pan, adding an equal amount of sugar. Cook over medium heat for one hour and 10 minutes, or until the jelling point is reached. Strain the lemon juice and the water in which the seeds have been soaking into the pan. Cook the mixture for 10 minutes more, stirring with a wooden spoon.

Pour the hot marmalade into jars, making sure that the fruit and syrup are evenly distributed. Process *(pages 30-31)*.

RAYMOND ARMISEN AND ANDRÉ MARTIN
LES RECETTES DE LA TABLE NIÇOISE

Clear Marmalade

For a marmalade with a more complex flavor and a wider spectrum of peel colors, you can substitute one grapefruit, four limes and two sweet oranges for the six bitter oranges, as shown on pages 50-51.

To make about 7 ½ pints [3 ¾ liters]

6	large bitter oranges, thinly sliced, seeds removed and reserved	6
2	lemons, thinly sliced, seeds removed and reserved	2
3½ quarts	water	3½ liters
10 cups	sugar	2½ liters

Pour 3 quarts [3 liters] of the water over the sliced fruit. Put the seeds into a bowl and cover with the remaining water. Let all soak for 24 hours.

Put the fruit into a heavy enameled, tinned or stainless-steel pan. Strain the water from the seeds into the pan, tie the seeds in a cheesecloth bag and add them to the pan. Bring to a boil and simmer gently for two hours, then remove the bag of seeds and add the sugar. Stir over very low heat until the sugar dissolves, then bring to a boil, and boil rapidly for 30 minutes, or until a little marmalade, tested on a saucer, jells. Put into jars. Cover and process *(pages 30-31)*.

THE DAILY TELEGRAPH
NEW DISHES FROM THE DAILY TELEGRAPH

Orange Pectin

The white pith that separates the colored surface —or zest— of orange peel from the fruit is too bitter to use for grated peel, and most cooks discard it. However, pith reserved after the zest has been removed for another use can be cooked this way to yield valuable pectin for jams and jellies.

To make about 3 cups [¾ liter]

¼ lb.	white orange pith (about ¾ cup [175 ml.])	125 g.
2 tbsp.	fresh lemon juice	30 ml.
3½ cups	cold water	875 ml.

Grind the pith in a food grinder, and mix it with the lemon juice and 1 cup [¼ liter] of cold water. Let it stand for four hours, add the remaining water, and let it stand overnight. In the morning bring the liquid to a boil, and boil for 10 minutes. Cool and strain the juice through a jelly bag. Process in jars *(pages 30-31)* or freeze.

ANN SERANNE
THE COMPLETE BOOK OF HOME PRESERVING

Jams and Conserves

Annfield Plain Winter Jam

To make about 7 pints [3 ½ liters]

6	apples, peeled, cored and cut into chunks	6
4	lemons, very thinly sliced and seeded	4
6	bananas, peeled and sliced	6
4	oranges, peeled, cut into chunks and seeded	4
5 quarts	water	5 liters
	sugar	

Put all the fruit into a pan with the water, and simmer for two hours. Measure the fruit, and add 1½ cups [375 ml.] of sugar for each 2½ cups [625 ml.] of fruit. Warm the sugar in a 325° F. [150° C.] oven for a few minutes. Add the hot sugar to the fruit, stir well and boil hard to the jelling point. Stir the mixture well and put into jars. Process *(pages 30-31)*.

PEGGY HUTCHINSON
GRANDMA'S PRESERVING SECRETS

Mixed Fruit Jam

Marmelada od Raznog Voca

Because the fruits in this recipe are puréed after cooking, it is not strictly necessary to peel them first.

To make about 8 pints [4 liters]

2 lb. each	apples, pears, peaches and plums, peeled, cored or pitted, and sliced (about 2 quarts [2 liters] each)	1 kg. each
8	medium-sized tomatoes, peeled, seeded and sliced (about 1 quart [1 liter])	8
2 lb.	large grapes, halved and seeded (about 1½ quarts [1½ liters])	1 kg.
2 lb.	cantaloupe, casaba or Cranshaw, peeled, halved, seeded and sliced (about 1½ quarts [1½ liters])	1 kg.
4 cups	sugar	1 liter

Place all of the fruits except the melon in a large enameled, tinned or stainless-steel pan and cook, without water, over low heat for 30 to 40 minutes, or until they are completely tender. Purée the cooked fruits by pressing them through a strainer or food mill into a large bowl.

Purée the melon, then add it to the cooked fruit purée. Add the sugar. Return the mixture to the pan and cook over medium heat, stirring constantly with a wooden spatula. The jam is ready when the spatula leaves a clean path that closes up slowly on the bottom of the pan.

Spoon the jam into jars. Cover and process *(pages 30-31)*.

SPASENIJA-PATA MARKOVIC (EDITOR)
VELIKI NARODNI KUVAR

Simple Apricot Jam

Marmellata d'Albicocche alla "sans Façon"

To make about 5 pints [2 ½ liters]

6 lb.	slightly underripe apricots, peeled, halved, pitted, a few of the pits reserved and the flesh thinly sliced (about 5 quarts [5 liters])	3 kg.
9 cups	sugar	2¼ liters

Crack a few of the apricot pits, remove and skin the kernels, and chop them very fine. Place the apricots, chopped kernels and sugar in an enameled, tinned or stainless-steel pan. Cook over medium heat, stirring frequently. Occasionally mash the pieces of apricot that have not yet broken up. When the jam is thick and sticky, after approximately one hour, put it in jars. Process *(pages 30-31)*.

IL CUOCO PIEMONTESE RIDOTTO ALL'ULTIMO GUSTO

Hunza Apricot Jam

The Hunza are an Islamic tribe of northern Kashmir; the isolated, fruit-growing valleys they inhabit are sometimes compared to the fictitious paradise of Shangri-La.

Any dried fruit or combination of dried fruits may be prepared in this manner. Spoonfuls of this jam can be frozen, or mixed with chopped nuts and coconut, to be eaten as ice cream or sweetmeats.

To make about 1 cup [¼ liter]

16	dried apricots, cut fine with scissors or a knife	16
	tepid water	
1 tbsp.	lemon juice	15 ml.
2 tsp.	honey	10 ml.
¼ tsp.	almond extract	1 ml.
3 tbsp.	plain yogurt	45 ml.

Add enough tepid water to the apricots to just cover them. Let them stand overnight or up to 24 hours. Add the lemon juice, honey and almond extract to the apricots, and press the mixture into a thick paste with a spoon or your hand, or in a blender. Fold in the yogurt. The jam will keep, refrigerated, for about three weeks.

SALLY DEVORE AND THELMA WHITE
THE APPETITES OF MAN

Blackberry or Raspberry Jam

Mûres ou Ronce

The addition of lemon juice is not indispensable to the jelling of the jam if you add a few unripe berries.

To make about 2 ½ pints [1 ¼ liters]

2 to 2½ lb.	ripe blackberries or raspberries, with a few that are still unripe (about 3½ pints [1 ¾ liters])	1 kg.
3 cups	sugar	¾ liter
2 tbsp.	fresh lemon juice (optional)	30 ml.

Place the berries and sugar in a bowl in alternate layers, and let the berries macerate overnight at room temperature.

The next day, add the lemon juice, if using, and bring the mixture to a boil in a saucepan. Skim thoroughly, and cook until the jam reaches the jelling point. It is best to undercook this jam slightly because it will stiffen when cool. Put it into jars. Process *(pages 30-31)*.

HUGUETTE CASTIGNAC
LA CUISINE OCCITANE

Four-Fruit Jam

Confiture de Quatre Fruits

To make about 12 pints [6 liters]

4 lb.	cherries, pitted (about 4 quarts [4 liters])	2 kg.
3 lb.	strawberries, hulled (about 5 pints [2½ liters])	1½ kg.
3 lb.	currants, stemmed (about 5 pints [2½ liters])	1½ kg.
3 lb.	raspberries, hulled (about 6 pints [3 liters])	1½ kg.
	sugar	

Measure the prepared fruits separately, and measure a quantity of sugar equal to their combined volume. Pour into an enameled, tinned or stainless-steel pan ¾ cup [175 ml.] of water for each 2 cups [½ liter] of sugar. Add the sugar, mix well, and cook over low heat until the sugar dissolves. Increase the heat and boil the syrup for 10 minutes, or until it reaches a temperature of 235° F. [112° C.]. Add the cherries, return to a boil, then reduce the heat and simmer for 15 minutes. Add the strawberries, return to a boil, and then simmer for 15 minutes more. Add the currants and cook for five minutes. Finally, add the raspberries. Cook for about 10 minutes, then test for jelling. When the jelling point is reached, remove the pan from the heat. Skim the jam and put it into jars. Process *(pages 30-31)*.

MISETTE GODARD
LE TEMPS DES CONFITURES

Sunshine Preserves

To make about 1 ½ pints [¾ liter]

1 lb.	cherries, pitted (about 1 quart [1 liter])	½ kg.
1½ cups	honey	375 ml.
1½ cups	sugar	375 ml.

Combine the honey and sugar. Bring slowly to the boiling point, add the cherries and cook for 12 minutes. Pour the mixture out onto shallow dishes; cover with glass—raised slightly above the dishes so that liquids can evaporate—and allow the mixture to stand in the sunshine for one or two days. Seal in hot, sterilized glasses. Refrigerate.

AMERICAN HONEY INSTITUTE
OLD FAVORITE HONEY RECIPES

Chestnut Jam

Castañas en Dulce

To make about 4 pints [2 liters]

3 lb.	chestnuts (about 3 quarts [3 liters])	1½ kg.
about 6 cups	water	about 1½ liters
about 5 cups	sugar	about 1¼ liters
about 1 tsp.	vanilla extract	about 5 ml.

Cut a cross in the flat end of each chestnut, place the chestnuts in a large pan, and cover them generously with water. Parboil them for 10 minutes. A few at a time, drain, shell and peel the chestnuts. Discard the parboiling water. Return the peeled chestnuts to the pan, add 5 cups [1¼ liters] of fresh water, cover, and simmer for about 45 minutes until the chestnuts are tender. Chop them coarse and purée them through a food mill.

Measure the chestnut purée, then measure out an equal amount of sugar. Place the sugar in a pan and, for each cup [¼ liter] of sugar, add ¾ cup [175 ml.] of water. Bring this syrup mixture to a boil, stirring until the sugar dissolves, then add the chestnut purée and a little vanilla extract. Cook over low heat for about 15 minutes. Pack the jam in jars, and process *(pages 30-31)*.

LUIS RIPOLL
COCINA DE LAS BALEARES

Elderberry and Plum Jam

This delightful jam has the flavor of Burgundy wine.

	To make about 5 pints [2 ½ liters]	
2½ lb.	elderberries, stemmed (about 7 cups [1 ¾ liters])	1 ¼ kg.
2 lb.	barely ripe, dark-skinned plums, slit open (about 1 ½ quarts [1 ½ liters])	1 kg.
2 cups	water	½ liter
8 cups	sugar	2 liters

Bring the elderberries to a boil with half of the water. Cook to a pulp—approximately 15 minutes—and strain overnight through a jelly bag. The elderberry pulp will render about 1 quart [1 liter] of juice.

Next day, cook the plums with the rest of the water until the pits come to the top; then remove the pits and return some of the peeled kernels to the plums. When the plums are quite soft, add the elderberry juice and the sugar. Stir until the sugar is dissolved. Bring the mixture to a boil and cook rapidly until the jam sets when tested. Pour into warm jars and cover. Process *(pages 30-31)*.

GERTRUDE MANN
BERRY COOKING

Gooseberry Jam

Marmolada z Agrestu

To make gooseberry and pumpkin jam, combine the finished gooseberry jam with an equal amount of pumpkin pulp, made by boiling peeled and seeded pumpkin pieces until tender, puréeing the flesh, and cooking it for 20 minutes, or until it is thick. After combining the pumpkin with the gooseberry jam, simmer the mixture for 10 minutes or so.

Another alternative is to add three parts of cooked and puréed pumpkin and two parts of sieved raw raspberries to each five parts of gooseberry jam. Simmer the mixture for a few minutes, or until thickened. This produces a jam with a wonderful aroma.

	To make about 4 pints [2 liters]	
4 lb.	young gooseberries, chopped (about 3 quarts [3 liters])	2 kg.
8 cups	sugar	2 liters

Cook the gooseberry pulp over low heat for about 45 minutes, or until it is reduced to half its original volume. Add the sugar, and simmer, stirring constantly, until the mixture thickens to the consistency of jam. While still hot, pour into jars. Process *(pages 30-31)*.

Z. CZERNY AND M. STRASBURGER
ZWIENIE RODZINY

Gooseberry and Raspberry Jam

	To make about 10 pints [5 liters]	
3 lb.	gooseberries, cleaned (about 2½ quarts [2½ liters])	1½ kg.
6 lb.	raspberries (about 12 pints [6 liters])	3 kg.
	sugar	

Place the gooseberries in an enameled, tinned or stainless-steel pan. Cover with cold water. Bring slowly to a boil. Simmer gently for about one hour, until the berries are softened into a mush, then pour into a jelly bag. Leave the berries in the bag without squeezing or mashing until all the juice is extracted, then measure.

Pour the juice into a preserving pan, adding 1 cup [¼ liter] of sugar for each 1¼ cups [375 ml.] of gooseberry juice, plus 9 cups [2¼ liters] of extra sugar to sweeten the raspberries. Stir gently over low heat until the sugar is dissolved, then bring to a boil. Add the raspberries. Return to a boil, and boil for about 15 minutes, until the jam jells when tested. Pour into jars and cover. Process *(pages 30-31)*.

ELIZABETH CRAIG
THE SCOTTISH COOKERY BOOK

Spiced Grapes

	To make 6 pints [3 liters]	
8 lb.	grapes, picked over (about 5 quarts [5 liters])	3½ kg.
1 quart	water	1 liter
	sugar	
1 ½ tsp.	ground ginger	7 ml.
1 ½ tsp.	ground cloves	7 ml.
1 ½ tsp.	ground cinnamon	7 ml.
1 ½ tsp.	grated nutmeg	7 ml.

Boil the grapes fast in the water for 10 minutes, or until they are soft enough to purée through a sieve. Measure the pulp, and add an equal measure of sugar if the grapes are green, or at the rate of three cupfuls [¾ liter] to every four [1 liter] of pulp for ripe grapes. Boil the pulp and sugar until quite thick, about 20 minutes, then add the spices. Let simmer and stir for a minute or two until the spices are well blended; then pour into jars and cover. Process *(pages 30-31)*.

MAY BYRON
MAY BYRON'S JAM BOOK

Grape and Pear Jam

Le Raisiné de Ménage

Windfall or bruised pears may be used for this recipe as long as all the damaged parts are removed.

To obtain the grape juice, crush about 6 pounds [2¾ kg.] of grapes (black, white, or a mixture of the two). Bring the crushed grapes quickly to a boil in a preserving pan, then pour them into a jelly bag or a cheesecloth-lined sieve. Drain, pressing the cloth to extract all the juice.

To make about 6 pints [3 liters]

2 quarts	grape juice	2 liters
4 lb.	pears, peeled, cored and quartered (about 3 quarts [3 liters])	2 kg.
2 cups	sugar	½ liter

Pour the grape juice into an enameled, tinned or stainless-steel pan, and add the pears. Cook very gently until the pears are tender, approximately 20 minutes. Remove them with a slotted spoon. Add the sugar to the juice in the pan, and cook, skimming occasionally, over medium heat until the mixture has reduced to a thick syrup. Return the pears to the pan, and cook, stirring, until the pears have reduced to a pulp and the jam is thick—about 15 minutes. Pour into jars. Process *(pages 30-31)*.

LE CORDON BLEU

Old-fashioned Wild Grape Conserve

To make about 8 cups [2 liters]

3 lb.	ripe wild grapes (about 2 quarts [2 liters])	1½ kg.
2 cups	raisins	½ liter
2-inch	strip fresh orange peel	5-cm.
6 to 8 tbsp.	strained fresh orange juice	90 to 120 ml.
2½ cups	sugar	625 ml.
1 cup	finely chopped hickory nuts or pecans	¼ liter

Slip the skins off the grapes. Cook the naked fruit for a few minutes, then put through a food mill or a sieve to remove the seeds. Grind the raw skins in a food grinder with the raisins and the orange peel. Combine the seeded pulp and the skin-raisin mixture, and add the orange juice and sugar. Cook very slowly for about an hour, until the mixture thickens. Then add the nut meats, stir well, and cook for five minutes more. Pour into jars and cover. Process *(pages 30-31)*.

EUELL GIBBONS
STALKING THE WILD ASPARAGUS

Papaya Marmalade

To make about 7 cups [1 ¾ liters]

10 cups	sliced firm, ripe papaya	2½ liters
1 cup	shredded fresh pineapple	¼ liter
2 tbsp.	grated orange peel	30 ml.
2 tbsp.	grated lemon peel	30 ml.
½ cup	fresh orange juice	125 ml.
½ cup	fresh lemon juice	125 ml.
3 tbsp.	grated fresh ginger	45 ml.
½ tsp.	salt	2 ml.
	sugar	

Combine all of the ingredients except the sugar. Boil for 30 minutes, until the papaya and pineapple are tender. Measure the cooked fruit and add an equal amount of sugar. Cook together for 30 minutes, stirring frequently. Pour the mixture into hot jars, cover and process *(pages 30-31)*.

CAREY D. MILLER, KATHERINE BAZORE, MARY BARTOW
FRUITS OF HAWAII

Peach Preserves

To make about 4 pints [2 liters]

4 lb.	firm, ripe peaches, scalded, peeled, pitted and chopped (about 7 cups [1 ¾ liters])	2 kg.
1	large lemon or 2 small lemons, finely chopped	1
1	small orange, finely chopped	1
5 cups	sugar	1¼ liters
½ tsp.	ground ginger or ¼ cup [50 ml.] chopped candied ginger	2 ml.
½ cup	almonds, blanched and slivered	125 ml.

Combine the lemons, orange and peaches in an enameled, tinned or stainless-steel pan, and simmer gently for 15 to 20 minutes, until the orange and lemon peels are tender. Add the sugar and ginger, and bring to a boil, stirring until the sugar is dissolved. Boil rapidly until the mixture sheets off a spoon or registers 222° F. [106° C.] on a candy thermometer. Add the almonds during the last five minutes of cooking.

Pour into hot jars and cover. Process *(pages 30-31)*.

JEAN HEWITT
NEW YORK TIMES NEW ENGLAND HERITAGE COOKBOOK

Peach Jam

For spiced peach jam, make a spice bag by wrapping 1 teaspoon [5 ml.] of whole cloves, ½ teaspoon [2 ml.] of whole allspice and one cinnamon stick in cheesecloth. Add the bag to the peaches when you add the water. Remove the bag before pouring the jam into jars.

To make about 4 pints [2 liters]

2 lb.	ripe peaches, peeled and pitted (about 1½ quarts [1½ liters])	1 kg.
½ cup	water	125 ml.
6 cups	sugar	1½ liters

In a heavy enameled, tinned or stainless-steel pan, crush the peaches with a potato masher or pestle. Add the water. Cook gently for 10 minutes, then add the sugar. Stirring occasionally until the sugar dissolves, slowly bring the mixture to a boil. Stirring often to prevent sticking, cook rapidly until the mixture is thick, about 15 minutes. Pour the jam, boiling hot, into hot jars, leaving ½ inch [1 cm.] headspace. Adjust the caps. Process for 15 minutes in a boiling water bath.

THE BALL BLUE BOOK

Peach, Orange and Walnut Conserve

To make about 12 pints [6 liters]

9 lb.	yellow peaches, scalded, peeled, halved, pitted and sliced (about 6 quarts [6 liters])	4 kg.
6	oranges, the peel grated, the pith removed, the flesh chopped and the juice reserved	6
1 cup	shelled walnuts, chopped	¼ liter
5	lemons, the peel grated, the pith removed, the flesh chopped and the juice reserved	5
⅔ cup	water	150 ml.
18 cups	sugar	4½ liters

Place the peaches in an enameled, tinned or stainless-steel pan with the grated orange and lemon peels and the water. Stirring occasionally, cook over low heat for 25 to 30 minutes, until the peaches are tender. Set the sugar to warm slightly in a 300° F. [150° C.] oven.

When the peaches are tender, add the chopped flesh and the juice of the oranges and lemons. Stir in the warmed sugar and continue stirring until all the sugar has dissolved. Then add the chopped walnuts and boil rapidly until the jelling point is reached—about 12 minutes. Ladle the jam into warm, dry jars and cover. Process *(pages 30-31)*.

PETITS PROPOS CULINAIRES III

Hodgepodge

To make about 20 pints [10 liters]

3 lb. each	peaches, quinces, pears and apples, peeled, quartered, and pitted or cored	1½ kg. each
about 6 cups	watermelon rind, cut into pieces	about 1½ liters
about 2 cups	cantaloupe rind, cut into pieces	about ½ liter
5	lemons, peeled, sliced and seeded	5
½	pineapple, peeled, cored and cut into chunks	½
	brown sugar	

Measure the prepared fruit, and for every 12 quarts [12 liters] of fruit add 6½ quarts [6½ liters] of sugar. Put the fruit and sugar into an enameled, tinned or stainless-steel pan, and boil, stirring occasionally, for 40 minutes, or until the quinces are well cooked and tender. Continue to boil for about 20 minutes, or until the fruit jells. Put the mixture into widemouthed preserving jars so that it can be unmolded when needed for the table. Cover and process *(pages 30-31)*.

MISS TYSON
THE QUEEN OF THE KITCHEN

Gingered Pear Jam

To make about 5 pints [2½ liters]

12	ripe pears, peeled, cored and cubed	12
	sugar	
3	pieces preserved ginger, finely chopped, with ⅓ cup [75 ml.] of the syrup from the jar	3
6 tbsp.	strained fresh lemon juice	90 ml.
about 1 cup	water	about ¼ liter

Measure the cubed pears, and measure an equal amount of sugar. Combine in an enameled, tinned or stainless-steel pan, the ginger, the syrup from the ginger jar, the lemon juice and the water. Bring slowly to a boil, stirring often while the juice from the pears dissolves the sugar; ripe pears should make enough liquid to cover—but, if necessary, gradually add more water. Cook at a rolling boil for about 20 minutes, stirring from time to time. When the mixture reaches the jelling point, remove it from the heat, cool it a little, and skim it very carefully. Pour the jam into jars, cap and process *(pages 30-31)*.

JULIETTE ELKON & ELAINE ROSS
MENUS FOR ENTERTAINING

Pyrenean Pear Jam

Poires Pyrénéennes

To make about 1 ½ pints [¾ liter]

2 to 2½ lb.	pears, peeled (about 1 ½ quarts [1 ½ liters])	1 kg.
about 1⅓ cups	sugar	about 325 ml.
4-inch	strip dried orange peel, pounded to a powder	10-cm.
1 tsp.	vanilla extract	5 ml.

Cook the pears in about ½ inch [1 cm.] of water, over low heat, for about 15 minutes, or until they are soft. Purée the pears through a sieve or food mill. Measure the pear purée, and add ⅓ cup [75 ml.] of sugar for each cup [¼ liter] of purée. Add the pounded orange peel and the vanilla extract. Stirring constantly, cook the mixture until it thickens, approximately 45 minutes. Put into jars. Process *(pages 30-31)*.

HUGUETTE CASTIGNAC
LA CUISINE OCCITANE

Red Pear Jelly

Gelée Rouge de Poire

Strictly speaking, this is not a jelly since it uses the pulp of the pears as well as the juice. However, the pulp dissolves completely. My choice of wine is a red Bordeaux.

To make about 10 pints [5 liters]

6 to 7 lb.	pears, peeled, quartered and cored (about 5 quarts [5 liters])	3 kg.
about 2 quarts	red wine	about 2 liters
	sugar	

Put the pears in a saucepan and add just enough wine to cover them. Bring to a boil, cover, and cook over low heat until the pears are soft—from 10 to 20 minutes, depending on the variety and ripeness of the pears.

Purée the pear mixture through a very fine sieve or food mill into the bowl. Measure the pear purée. Measure an equal quantity of sugar.

Pour into an enameled, tinned or stainless-steel pan ¾ cup [175 ml.] of water for each 2 cups [½ liter] of sugar. Add the sugar, mix well, and cook over low heat until the sugar is completely dissolved. Increase the heat, and cook until the syrup reaches the soft-ball stage—235° F. [112° C.] on a candy thermometer. Add the pear purée and mix well. Stirring constantly, cook over medium heat until the jelling point is reached, about 15 minutes. Pour the jam into jars and cover. Process *(pages 30-31)*.

MISETTE GODARD
LE TEMPS DES CONFITURES

Harrods Creek Fall Fruit Conserve

This conserve can be served with game or roasts. It is also delicious used as a topping for ice cream if a little brandy, whisky or rum, or a combination of two of them, is added before serving. To 1 cup [¼ liter] of conserve, add 2 to 3 tablespoons [30 to 45 ml.] of liquor, or more to taste.

To make about 9 pints [4 ½ liters]

5 lb.	firm pears, peeled, cored and cut into slivers (about 4 quarts [4 liters])	2½ kg.
1	grapefruit, unpeeled, but thinly sliced and seeded	1
1	orange, unpeeled, but thinly sliced and seeded	1
3	lemons, unpeeled, but thinly sliced and seeded	3
10 cups	sugar	2½ liters
4 cups	dark seedless raisins	1 liter
2 cups	pecans or walnuts	½ liter

Put the pears, grapefruit, orange and lemons through a food grinder, saving the juices. Combine the juices, fruits and the sugar, and mix well. Let stand in a covered bowl overnight.

The next morning, put the fruits and sugar in an enameled, tinned or stainless-steel pan. Let come to a boil, then reduce the heat to very low and cook for 45 minutes, stirring occasionally to prevent sticking. Add the raisins and cook for 45 minutes longer. Add the nuts and cook one minute longer. Pour the conserve into jars, close and process *(pages 30-31)*.

MARION FLEXNER
OUT OF KENTUCKY KITCHENS

Persimmon Jelly

To make about 8 cups [2 liters]

6 lb.	ripe persimmons	3 kg.
1 cup	water	¼ liter
4 cups	sugar	1 liter
1	lemon, the peel only, cut into strips	1

Wash and stem the persimmons. Halve them, remove all of the pulp from the halves, and press it through a strainer or food mill. Measure the puréed pulp; there should be about 4 cups [1 liter]. Boil the water, sugar and lemon peel until the mixture reaches the soft-ball stage, 235° F. [112° C.] on a candy thermometer. Remove the lemon peel and stir in the persimmon pulp. Cook over low heat until the mixture jells. Test by letting a drop fall on a dry plate; cool, and lift one edge of the plate—if the drop remains rounded, the jelly is ready. Pour the jelly into jars. Process *(pages 30-31)*.

ANGELO SORZIO
THE ART OF HOME CANNING

Pineapple and Apricot Conserve

To make 5 pints [2 ½ liters]

1	large pineapple, peeled, cored and cubed	1
4 lb.	apricots, scalded, peeled, quartered and pitted (about 2 quarts [2 liters])	2 kg.
1 cup	hot water	¼ liter
5 cups	sugar	1 ¼ liters

Add the water to the sugar in a stainless-steel, tinned or enameled pan and bring this syrup mixture to the boiling point. Add the pineapple to the syrup and cook over medium heat until tender, about 10 minutes. Then drop in the apricots and boil for several minutes, or until they are tender. Have hot jars ready, fill them with the conserve, close and process *(pages 30-31)*. Before storing, label each jar.

WOMAN'S INSTITUTE LIBRARY OF COOKERY

Plum Conserve

To make about 4 pints [2 liters]

3 lb.	plums, quartered and pitted (about 2 quarts [2 liters])	1 ½ kg.
2	thin-skinned oranges, halved, seeded and finely ground through a food grinder	2
8 cups	brown sugar	2 liters
4 cups	raisins	1 liter
1 cup	shelled walnuts, chopped	¼ liter

In a heavy saucepan, combine the plums, ground oranges, sugar and raisins. Stirring occasionally, cook until the mixture is very thick, about one and one half hours.

Stir in the nuts and cook for 20 minutes longer. Pour into hot jars. Process *(pages 30-31)*.

JEAN HEWITT
THE NEW YORK TIMES NATURAL FOODS COOKBOOK

Plum-Rum Conserve

To make about 2 pints [1 liter]

2 lb.	red plums, halved, pitted and sliced (about 6 cups [1 ½ liters])	1 kg.
1 cup	water	¼ liter
3 cups	sugar	¾ liter
1 cup	coarsely chopped walnuts	¼ liter
⅓ cup	dark rum	75 ml.

Combine the plums and the water in an enameled, tinned, stainless-steel or heatproof glass saucepan, and simmer for

three minutes. Add the sugar and return the plums to a simmer, stirring. Remove from the heat, cover with a towel, and let the mixture stand overnight, or from 12 to 24 hours.

Stirring now and then, boil the plums gently until the mixture begins to get thick and sticky. From then on, stir constantly. Boil until this conserve reaches 225° F. [110° C.] on a sugar thermometer. This usually takes about 30 minutes, but the time will vary with the juiciness of the plums. Remove the saucepan from the heat and stir in the chopped walnuts, then add the rum.

Ladle the conserve into hot jars, leaving ½ inch [1 cm.] of headspace. Cover and process *(pages 30-31)*. Store the jars for at least two weeks before serving the conserve.

HELEN WITTY AND ELIZABETH SCHNEIDER COLCHIE
BETTER THAN STORE-BOUGHT

Greengage Jam
Confiture de Reine-Claude

To make about 5 pints [2 ½ liters]

6 to 7 lb.	ripe greengage plums, halved and pitted (about 4 quarts [4 liters])	3 kg.
about 8 cups	sugar	about 2 liters

Measure the pitted fruit and add ½ cup [125 ml.] of sugar for each cup [¼ liter] of fruit. Mix the fruit and sugar in a bowl and let stand overnight. The next day, pour the contents of the bowl into an enameled, tinned or stainless-steel serving pan, and cook the jam until the jelling point is reached. Put in jars. Process *(pages 30-31)*.

Alternatively, drain the juice from the fruit and cook this syrup until it has reached the soft-ball stage—that is, until a drop of syrup placed in cold water becomes rubbery (235° F. [112° C.] on a candy thermometer). Add the fruit and continue to cook until it is tender and translucent and the syrup has reached the jelling point. Put in jars. Process. This method keeps the fruit more neatly intact.

LE CORDON BLEU

Quince Marmalade
Compote de Coings

To make 1 ½ pints [¾ liter]

2 to 2½ lb.	quinces, carefully wiped, thinly sliced, but not peeled or cored (about 5 cups [1 ¼ liters])	1 kg.
1	vanilla bean	1
1 cup	water	¼ liter
2½ cups	sugar	625 ml.
5 tbsp.	rum	75 ml.

Put the quinces in an enameled, tinned or stainless-steel pan with the vanilla bean, and add ¾ cup [175 ml.] of the water.

Stirring frequently with a wooden spoon, cook over low heat for about two hours, or until you have a soft pulp.

Remove the vanilla bean, and let the quinces cool until lukewarm. Purée them through a food mill into a saucepan.

In another pan, mix the sugar with the remaining water, and cook over very low heat until it forms a clear, very thick syrup. Add the syrup to the quince purée, mix carefully and, stirring frequently, cook over very low heat for about 45 minutes. Cool the mixture slightly, then add the rum. Put into jars. Process *(pages 30-31)*.

TINA CECCHINI
LES CONSERVES DE FRUITS ET LÉGUMES

———————◆———————

Red Raspberry-Apricot Jam

To make about 4 pints [2 liters]

4 lb.	fresh red raspberries (about 8 pints [4 liters])	2 kg.
1½ cups	dried apricots	375 ml.
1 cup	water	¼ liter
6 cups	sugar	1½ liters
½ tsp.	salt	2 ml.

Simmer the apricots in the water for 15 minutes. Add the raspberries, sugar and salt. Cook rapidly for 15 minutes, or until the jam is of the desired consistency. Ladle the jam into sterilized jars to within ½ inch [1 cm.] of the jar tops. Wipe the jar rims; adjust the lids. Process in a boiling water bath for 15 minutes. Remove the jars and complete the seals, unless the jars are self-sealing.

NELL B. NICHOLS AND KATHRYN LARSON (EDITORS)
FARM JOURNAL'S FREEZING & CANNING COOKBOOK

———————◆———————

Rhubarb and Fig Jam

To make about 9 pints [4½ liters]

7 lb.	rhubarb, leaves removed, stalks trimmed and cut into pieces (about 6 quarts [6 liters])	3 kg.
1 lb.	dried figs, cut into fine shreds	½ kg.
11 cups	sugar	2¾ liters
1 cup	mixed candied fruit peel, chopped	¼ liter

Mix the rhubarb, figs and sugar in an earthenware crock or a large jar. Cover and let stand all night. The next day, boil the mixture for at least an hour, or until very thick, and add to it, before it is taken off the heat, the candied peel. Pour the jam into warm jars and cover. Process *(pages 30-31)*.

MRS. C. F. LEYEL
THE COMPLETE JAM CUPBOARD

Rhubarb Jam

This recipe is from a collection by a 19th Century English noblewoman. The jam should be processed (pages 30-31).

To make 9 pints [4½ liters]

6 to 7 lb.	rhubarb, leaves removed, stalks trimmed and cut into 2- to 3-inch [5- to 8-cm.] lengths (about 6 quarts [6 liters])	3 kg.
12 cups	sugar	3 liters
1-inch	slice fresh ginger, crushed roughly with a kitchen mallet and tied in cheesecloth (optional)	2½-cm.

Put the rhubarb in a stoneware jar or a casserole, cover with the sugar, add the ginger if using, and let it lie some hours—use no water, as the rhubarb is so full of juice. Put the jar in a pan of hot water to boil from 30 minutes to one and a half hours, according to whether the rhubarb is old or not; stir all the time. Remove the ginger and put the jam in jars.

CATHERINE FRANCES FRERE (EDITOR)
THE COOKERY BOOK OF LADY CLARK OF TILLYPRONIE

———————◆———————

Rose-Hip Jam

Confiture de Cynorrhodons

A rose hip is the seed pod —or fruit —left at the top of the stem of a rose after the flower has withered.

This jam is made from the fruit of the wild or dog rose. For the hips to be at their most flavorful, they should be deep red. Gather them only after the first frosts. The jam is a runny purée which never actually jells. To make it without wine, macerate and cook the fruit in water. Add 1½ cups [375 ml.] of sugar for each 2 cups [½ liter] of purée, and cook the mixture for 30 minutes.

To make about 4 pints [2 liters]

4 lb.	rose hips, black tips removed, pods halved (about 3 quarts [3 liters])	2 kg.
	dry white wine	
	sugar	

Place the hips in a bowl and cover them with white wine. Let them macerate for a week in a cool place, stirring them every day with a wooden spatula.

Place the hips and wine in an enameled, stainless-steel or tinned pan and simmer for one hour, or until the fruit is soft and pulpy. Press the mixture through a fine sieve to remove all of the fruit's skin, filaments and seeds. Measure the purée and add the same amount of sugar. Cook, stirring constantly, until the mixture comes to a boil. Cook for 10 minutes, skim and ladle into jars. Process *(pages 30-31)*.

GINETTE MATHIOT
JE SAIS FAIRE LES CONSERVES

Rhubarb and Orange Jam

Confiture de Rhubarbe à l'Orange

To make about 2 ½ pints [1 ¼ liters]

2 lb.	rhubarb, leaves removed, stalks trimmed and cut into pieces (about 2 quarts [2 liters])	1 kg.
1	orange, the peel finely chopped and the flesh trimmed of pith, thinly sliced and seeds removed	1
⅔ cup	water	150 ml.
2 cups	sugar	½ liter
¼ cup	rum	50 ml.
1 tsp.	vanilla extract	5 ml.

Put the rhubarb in an enameled, tinned or stainless-steel pan with the water, cover, and cook gently for about 15 minutes, or until tender. Drain the rhubarb and purée it through a sieve or food mill into a saucepan. Add the sugar, chopped orange peel and orange slices. Boil for 15 minutes, then add the rum and the vanilla extract. Remove from the heat and pack the jam in jars. Process *(pages 30-31)*.

TINA CECCHINI
LES CONSERVES DE FRUITS ET LÉGUMES

Rose-Hip Marmalade

Marmelad ot Shipki

A rose hip is the seed pod —or fruit —left at the top of the stem of a wild rose after the flower has withered.

To make about 4 pints [2 liters]

10 lb.	late-autumn rose hips, black tips removed (about 7 ½ quarts [7 ½ liters])	4 ½ kg.
7 cups	sugar	1 ¾ liters

Place the rose hips in a large pan, cover with boiling water and boil until completely soft. Press the fruit through a colander to remove the seeds and skins, then force the pulp twice through a very fine-meshed sieve to remove as many of the seed filaments as possible.

Stirring with a wooden spatula, boil the pulp rapidly until it is quite thick. Remove the pulp from the heat, stir in the sugar until it is dissolved, then cook over high heat, stirring constantly, until the mixture is so thick that the spatula leaves a path on the bottom of the pan. Spoon the hot marmalade into jars; cover and process *(pages 30-31)*.

SONYA CHORTANOVA
NASHA KUCHNIYA

Strawberry Jam

Before hulling the berries, wash them by plunging them into a large bowl of water, swirling them rapidly and removing them immediately with widespread hands, fingers splayed, to a colander. Do this in several batches, changing the water for each new batch. Hull them, and spread them loosely on a tray to avoid crushing them.

To make about 5 pints [2 ½ liters]

11 lb.	strawberries, hulled (about 8 quarts [8 liters])	5 kg.
4 cups	sugar	1 liter
1 quart	water	1 liter

In an enameled, tinned or stainless-steel pan, prepare a syrup by cooking the sugar and water together over low heat until the sugar is dissolved. Boil for a minute over high heat.

Throw in about 3 cups [¾ liter] of the strawberries. Return the syrup to a boil, cook for a minute or so, and remove the strawberries with a slotted spoon to a colander placed over a large bowl. Boil the syrup until it has reduced to its former volume, and throw in another batch of strawberries.

Continue in this manner, removing each batch of berries to the colander after one minute at full boil, and reducing the syrup again after the removal of each batch. From time to time, drain the juices that collect beneath the colander back into the preserving pan. When the last batch has been removed, reduce the syrup once again and return all the strawberries to the pan at once, along with any more juices that have collected. Cook at a light boil for about 10 minutes, stirring gently from time to time. Cool slightly before ladling the jam into jars, being careful to distribute the fruit equally among the jars. Process *(pages 30-31)*.

PETITS PROPOS CULINAIRES V

Sun-kissed Strawberries

To make 2 pints [1 liter]

1 ½ lb.	strawberries (about 4 cups [1 liter])	¾ kg.
4 cups	sugar	1 liter
¼ cup	strained fresh lemon juice	50 ml.
¼ cup	water	50 ml.

Mix the sugar, lemon juice and water together. Heat slowly, stirring, and boil for four minutes. Add the berries, heat to boiling and boil for 10 minutes. Pour the mixture into shallow pans or platters, and cover it with sheets of glass, slightly raised to permit evaporation. Let the mixture stand in hot sunshine for three or four days, stirring occasionally, until the syrup is rich and thick and the berries are plump and translucent. Seal in small jars or glasses. Refrigerate.

GENEVIEVE CALLAHAN
THE NEW CALIFORNIA COOK BOOK

Seedless Strawberry Jam

Confettura di Fragole

To make about 4 pints [2 liters]

4 lb.	strawberries, hulled (about 3 quarts [3 liters])	2 kg.
3½ cups	sugar	875 ml.

Press the strawberries through a sieve, and add the sugar to the resulting purée. Let the mixture stand overnight. The next day, cook over medium heat, stirring often, for about 15 minutes, or until the jam thickens somewhat. Put the jam into jars, cover and process *(pages 30-31)*.

IL RE DEI CUOCHI

Strawberry and Pineapple Conserve

To make about 4 pints [2 liters]

3 lb.	strawberries, hulled (about 2 quarts [2 liters])	1½ kg.
1	medium-sized pineapple, trimmed, peeled, quartered, cored and cut into small pieces (about 2 cups [½ liter])	1
6 cups	sugar	1½ liters

Combine the pineapple and sugar, and simmer over low heat for 10 minutes. Add the strawberries, and cook slowly for about 15 minutes, or until the mixture is thick and clear. Pour into glasses or jars and cover. Process *(pages 30-31)*.

HELEN CORBITT
HELEN CORBITT'S COOK BOOK

Bean Jam

Adzuki beans are tiny, round dried red beans, available in Oriental groceries or health-food stores.

This is usually made from adzuki beans, but dried red kidney beans can be used just as effectively. For white jam, navy beans are substituted. If you use kidney or navy beans, soak them for six to eight hours before use. The jam is used as a filling for small steamed buns.

To make about 2½ pints [1¼ liters]

1 lb.	adzuki beans	½ kg.
2 tsp.	salt	10 ml.
2 cups	sugar	½ liter

Put the beans in a heavy pan, add just enough water to cover, and boil until the beans are soft and the skins have started to peel off, about one hour.

For smooth bean jam, drain the boiled beans, force them through a sieve or food mill, place the pulp in a clean towel or piece of cheesecloth, and squeeze out any remaining liquid. Return the pulp to the pan, add the salt and, over low heat, gradually stir in the sugar. Continue heating, stirring, and cook down to form a stiff, smooth mixture.

For coarse bean jam, drain the boiled beans, add the salt and sugar, and cook over low heat, mashing to a paste with a wooden spoon. Put the jam in glass jars with tight-fitting lids. Process *(pages 30-31)*.

DAVID SCOTT
THE JAPANESE COOKBOOK

Carrot Jam

Confitures de Carottes

To make about 5 pints [2½ liters]

2 lb.	tender young carrots, cut into thin strips (about 2 quarts [2 liters])	1 kg.
6 cups	sugar	1½ liters
3 or 4	lemons, the peel grated and the juice strained	3 or 4

In an enameled, tinned or stainless-steel pan, make a layer of carrots. Cover with some of the sugar and lemon peel, and sprinkle with a little lemon juice. Repeat these layers, finishing with a layer of sugar. Add enough water to cover. Cover the pan, and cook over very low heat for four hours. When the jelling point is reached, put the jam into jars, cover them and process *(pages 30-31)*.

ÉDOUARD NIGNON (EDITOR)
LE LIVRE DE CUISINE DE L'OUEST-ÉCLAIR

Country Jam

Confitures Bien de Chez Nous

To make about 1½ pints [¾ liter]

5 to 6	medium-sized carrots, grated	5 to 6
1	pear, peeled, cored and sliced	1
1 lb.	honey (1½ cups [375 ml.])	½ kg.
2 tbsp.	sugar	30 ml.
3 tbsp.	strained fresh lemon juice	45 ml.
¼ tsp.	ground cinnamon	1 ml.

Cook all of the ingredients together over medium heat for 30 to 40 minutes, until the carrots are transparent and the jelling point is reached. Put in jars. Process *(pages 30-31)*.

AMICALE DES CUISINIERS ET PÂTISSIERS AUVERGNATS DE PARIS
CUISINE D'AUVERGNE

Red Chili Jam

To make about 1 ½ pints [¾ liter]

12	large fresh medium-hot red chilies (preferably Anaheim chilies), stemmed, seeded and finely chopped	12
2	small lemons, quartered	2
½ cup	cider vinegar	125 ml.
3 cups	sugar	¾ liter

Combine the chilies, lemons and vinegar, and cook for approximately 30 minutes, or until the chilies are tender. Remove the lemon quarters and add the sugar. Boil for 10 minutes, or until the jam reaches 8° F. [5° C.] above boiling on a jelly thermometer, or passes the sheet test. Spoon the jam into jars, cover and process *(pages 30-31)*.

JANE BUTEL
JANE BUTEL'S TEX-MEX COOKBOOK

Eggplant Preserve

Confiture d'Aubergines

To make about 5 pints [2 ½ liters]

2 to 2 ½ lb.	pear-sized eggplants, stems left on, peeled, and pricked all over with a fork	1 kg.
4 cups	sugar	1 liter
2 lb.	honey	1 kg.
¼ cup	ground ginger	50 ml.

Put the eggplants in a large bowl and cover them with 2 quarts [2 liters] of cold water. Soak them for three days, changing the water every day.

Bring a large pan of water to a boil, plunge in the eggplants and cook them for 10 minutes. Rinse them in cold water and drain them thoroughly.

In a saucepan, mix the sugar with 2 cups [½ liter] of water. Add the eggplants and bring the mixture very slowly to a boil. Cook slowly for 30 minutes, then, over the lowest possible heat, cook for one hour more. When the syrup is quite thick and the eggplants are translucent, stir in the honey and ginger. Cook for 10 minutes and remove from the heat. Put into jars. Process *(pages 30-31)*.

IRÈNE ET LUCIENNE KARSENTY
LA CUISINE PIED-NOIR

Prickly Pear Preserves

To make 2 pints [1 liter]

4 lb.	prickly pears (about 2 ½ quarts [2 ½ liters])	2 kg.
1 ½ cups	sugar	375 ml.
⅝ cup	water	150 ml.
2 ½ tbsp.	fresh lemon juice	37 ml.
1	slice orange, ¼ inch [6 mm.] thick	1

Rub off the bristles with paper towels, wash the pears, peel them, cut them into halves and remove the seeds. Combine with the other ingredients, and cook until the pears are transparent. Remove the orange slice. Ladle into hot jars and cover. Process *(pages 30-31)*.

STANLEY SCHULER AND ELIZABETH MERIWETHER SCHULER
PRESERVING THE FRUITS OF THE EARTH

Pumpkin and Dried Apricot Jam

Confiture de Potiron et d'Abricots Secs

Winter squashes such as Hubbard, butternut or acorn can be substituted for the pumpkin.

To make about 10 pints [5 liters]

6 lb.	pumpkin, quartered, seeded, peeled, and cubed (about 4 ½ quarts [4 ½ liters])	3 kg.
2 lb.	dried apricots, cut into strips	1 kg.
1 ½ quarts	water	1 ½ liters
11 cups	sugar	2 ¾ liters

Soak the apricots in the water for 24 hours. Drain, reserving the water. Cook the pumpkin in the apricot water for 30 minutes, or until the pumpkin is soft. Purée the pumpkin through a food mill. Add the sugar, and cook the mixture for 30 minutes. Finally, add the apricots and cook for 30 minutes more, or until the apricots are soft and the jam has thickened. Put in jars. Process *(pages 30-31)*.

GINETTE MATHIOT
JE SAIS FAIRE LES CONSERVES

Balkan Pumpkin Jam

Sezonen Marmelad ot Tikvi

To prepare the baked pulp, you will need 10 pounds [4 ½ kg.] of whole pumpkin, or you may substitute a like amount of Hubbard or butternut squash. Halve the pumpkins or squashes crosswise. Scoop out the seeds and fiber, place the halves, cut sides up, on a baking sheet, and bake in a 400° F. [200° C.] oven for about three hours, or until the flesh is tender when pierced with a fork or skewer. Scoop out the flesh and press it through a food mill.

During the winter months, this jam can be served as a dessert, mixed with crushed walnuts or roasted blanched almonds. It can be used to make lovely pumpkin pies and strudels—much nicer than the common apple strudel.

To make about 5 pints [2 ½ liters]

2½ quarts	baked pumpkin pulp	2½ liters
4 cups	sugar	1 liter
½	vanilla bean, split lengthwise, or 1 tsp. [5 ml.] finely grated orange peel, or ½ tsp. [2 ml.] ground cinnamon	½

Place the pumpkin pulp in a pan, and add the sugar and the vanilla bean or orange peel (cinnamon is added later). Stir over low heat for about 10 minutes, or until the sugar is dissolved. Increase the heat and boil, stirring with a long-handled wooden spatula, until the mixture turns a rich orange brown and the spatula leaves a clean path on the bottom of the pan, about 30 minutes. Remove from the heat and stir in the cinnamon if using; discard the vanilla bean. Pack into hot jars. Process *(pages 30-31)*.

LIDIYA KURDZHIEVA (EDITOR)
DOMASHNO PRIGOTVYANE NA ZIMNINA

Angel's Hair

Cabello de Ángel

Cabello de ángel is popular in Spain and is often used for pastry fillings.

To make about 3 pints [1 ½ liters]

2¼ lb.	winter squash, halved, seeded, sliced and peeled	1 kg.
1 quart	water	1 liter
4 cups	sugar	1 liter
1	lemon, quartered and seeded	1
1	cinnamon stick	1

In an enameled, tinned or stainless-steel pan, combine the squash flesh with the water, cover, and cook over low heat for about 30 minutes, or until the squash is soft. Drain, reserving the cooking liquid, and put the squash on a plate. Using two forks, pull the fibrous flesh apart and discard any seeds. Put the squash into a colander to drain thoroughly.

Mix the cooking liquid with the sugar and bring to a boil. Add the lemon and cinnamon. Boil for 15 minutes, remove the lemon and cinnamon, and add the squash. Stirring often, cook over medium heat for 30 minutes, or until the jelling point is reached. Put into jars. Process *(pages 30-31)*.

VICTORIA SERRA
TIA VICTORIA'S SPANISH KITCHEN

Sweet Potato Jam

Mermalada de Batata

When the sweet potatoes are cooked and peeled, they lose about a fifth of their weight. Peaches and pumpkins may be preserved in the same fashion.

To make about 3 pints [1 ½ liters]

2½ lb.	sweet potatoes	1¼ kg.
about 4 cups	sugar	about 1 liter

Place the sweet potatoes in a pot, cover with water, bring to a boil and cook until tender. Drain them and cool to tepid. Peel the sweet potatoes and purée them through a sieve. Measure the purée and combine it with an equal amount of sugar. Cook the mixture over very low heat, stirring with a wooden spatula to keep it from sticking or boiling over. In about 10 minutes, the mixture will be a thin, almost transparent jam. Pour it into jars, cover and process *(pages 30-31)*.

JOSÉ SARRAU
MI RECETARIO DE COCINA

Green-Tomato Jam

Tomates Vertes

To make about 2 ½ pints [1 ¼ liters]

2 lb.	green tomatoes, sliced (about 4 cups [1 liter])	1 kg.
3 cups	sugar	¾ liter
1 tsp.	vanilla extract or 2 tbsp. [30 ml.] fresh lemon juice	5 ml.

Layer the tomato slices with the sugar, add the vanilla extract or lemon juice, and let the tomatoes macerate overnight. The next day, cook the jam over medium heat for approximately 30 minutes, or until it reaches the jelling stage. Be careful: The jam cooks very quickly and will harden in the jars if it is overcooked; it is better to undercook the jam slightly. Put in jars. Process *(pages 30-31)*.

HUGUETTE CASTIGNAC
LA CUISINE OCCITANE

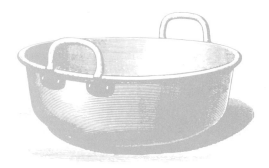

Tomato Marmalade

Mermelada de Tomate

To make about 2 pints [1 liter]

1 ½ lb.	large ripe tomatoes, peeled, halved and seeded	¾ kg.
2½ cups	sugar	625 ml.
1 cup	water	¼ liter
6 tbsp.	strained fresh lemon juice	90 ml.

Boil the sugar and water together for about five minutes to make a thick syrup. Stir in the tomatoes and the lemon juice. Stirring frequently, cook over medium heat for about 45 minutes, or until the tomatoes are translucent and the syrup thick. Put in jars and cover. Process *(pages 30-31)*.

VICTORIA SERRA
TIA VICTORIA'S SPANISH KITCHEN

Zucchini Jam

Confiture de Courges

To make about 6 pints [3 liters]

6 lb.	large zucchini, peeled, seeded and cubed (about 4 quarts [4 liters])	3 kg.
1 lb.	brown sugar	½ kg.
about 8 cups	granulated sugar	about 2 liters
¼ lb.	fresh ginger, crushed slightly with a mallet and tied in a cheesecloth bag	125 g.
1	lemon, peel only, pared in thin strips	1
⅓ cup	chopped mixed candied fruit peel	75 ml.

In a bowl, sprinkle the cubes of zucchini with the brown sugar and let the mixture stand for a day.

The following day, drain off the liquid from the zucchini. Put the liquid in a large pan and add 3 cups [¾ liter] of granulated sugar to each 2½ cups [625 ml.] of liquid. Put in the ginger, lemon peel and candied peel. Boil all this fast for 10 minutes, then add the zucchini cubes and cook slowly for three hours. Remove the lemon peel and ginger before putting the jam in jars. Process *(pages 30-31)*.

X. MARCEL BOULESTIN
A SECOND HELPING OR MORE DISHES FOR ENGLISH HOMES

Curds, Butters and Pastes

Blackberry Curd

To make about 1 ½ pints [¾ liter]

¾ lb.	blackberries (about 2½ cups [625 ml.])	350 g.
1	large cooking apple, peeled, cored and chopped	1
3 tbsp.	strained fresh lemon juice	45 ml.
8 tbsp.	unsalted butter	120 ml.
2 cups	sugar	½ liter
4	eggs, beaten	4

Put the blackberries and apple chunks into an enameled, tinned or stainless-steel pan without water to cook very slowly, so that the fruit is softened and the juice drawn out. Then purée the mixture through a strainer or food mill, and put the pulp into the top of a double boiler. Add the lemon juice, butter and sugar. Stirring occasionally, simmer very gently over hot—not boiling—water until the sugar has dissolved. Then gradually add the beaten eggs and continue to simmer until the mixture thickens. Put in jars and cover. Refrigerated, this curd will keep for two to three months.

DAVID & ROSE MABEY
JAMS, PICKLES & CHUTNEYS

Gooseberry Curd

Gooseberry curd is used as a filling for tarts and cakes.

To make about 1 ½ pints [¾ liter]

2 lb.	gooseberries, trimmed (about 6 cups [1½ liters])	1 kg.
¼ cup	water	50 ml.
4 tbsp.	butter	60 ml.
3	eggs, lightly beaten	3
2 cups	sugar	½ liter

In a large saucepan, bring the gooseberries and the water to a boil over high heat. Reduce the heat to low, cover the pan and simmer for 20 to 25 minutes, or until the gooseberries are soft and mushy. Remove the pan from the heat and push the fruit through a strainer into a medium-sized mixing bowl, pressing down on the fruit with a wooden spoon. Discard the skins left in the strainer.

In another bowl set over a pan of simmering water, melt the butter. Stir in the lightly beaten eggs, sugar and gooseberry purée. Cook the mixture, stirring frequently, for 25 to 30 minutes, or until it thickens.

Remove the bowl from the heat and pour the curd into clean, dry jam jars. Cover and process as for jam *(pages 30-31)*. Label the jars and keep them in a cool, dry place. The curd will last for three to four months.

MARYE CAMERON-SMITH
THE COMPLETE BOOK OF PRESERVING

My Lemon Curd

Lemon curd is used as a filling for tarts or cakes.

To make about 1 pint [½ liter]		
2	lemons, the peel grated, the juice strained	2
½ cup	superfine sugar	125 ml.
½ cup	granulated sugar	125 ml.
4 tbsp.	unsalted butter	60 ml.
3	eggs, beaten	3

Place the sugars in the top of a double boiler. Add the butter and the lemon peel and juice, then stir in the eggs. When blended, place over hot—not boiling—water. Stirring constantly, cook until the sugar dissolves and the mixture thickens, about 10 minutes. Do not allow to boil. Put in jars and cover at once. Refrigerated, this curd will keep for two to three months.

ELIZABETH CRAIG
THE SCOTTISH COOKERY BOOK

Rhineland Apple Butter

Rheinisches Apfelkraut

This apple butter should have the consistency of honey.

To make 1 pint [½ liter]		
11 lb.	apples, quartered (about 10 quarts [10 liters])	5 kg.
2 cups	water	½ liter

Cook the apples in a steamer until they dissolve into a purée, or simmer them with the water until soft, about 40 minutes.

Pour the apples into a jelly bag and let the juice drip through for a whole day. You will have about 3 quarts [3 liters] of juice. Put the juice in a large, heavy pan and bring it quickly to a boil. Cook uncovered until the juice has reduced to about 2 cups [½ liter] and is dark brown, stirring constantly toward the end of the cooking time. It will take about one hour and 40 minutes. Put the finished apple butter in a jar and cover. Process *(pages 30-31)*.

EIKE LINNICH
DAS GROSSE EINMACHBUCH

Apple Butter in the Oven

This easy recipe cooks into rich brown butter without plopping all over the stove and without burning.

To make about 5 pints [2½ liters]		
10 lb.	apples, quartered (about 9 quarts [9 liters])	4½ kg.
1¼ cups	water	300 ml.
6 cups	apple juice or sweet cider	1½ liters
1 tbsp.	ground cinnamon	15 ml.
1 tbsp.	ground cloves	15 ml.
1 tbsp.	grated nutmeg	15 ml.
10 cups	sugar	2½ liters

Cook the apples gently with the water until they are very soft, then force them through a food mill to make a purée. Pour the purée into a deep enameled, tinned or stainless-steel pan, and stir in the apple juice or cider, the spices and sugar. Cover the pan and put it into an oven preheated to 350° F. [180° C.]. Bake, stirring occasionally, until the mixture boils. Once it is boiling, turn the heat down to 250° F. [120° C.] and bake for five hours, or until the butter is stiff. Alternatively, turn the heat down to 200° F. [100° C.] after the mixture boils, and leave the apple butter to bake all night. In the morning, ladle the apple butter into jars and cap them. Process *(pages 30-31)*.

GRACE FIRTH
A NATURAL YEAR

Pennsylvania Dutch Apple Butter

Apple butter is much like a 19th Century English "apple marmalade"; in both cases the apples are cooked in cider, pulped and mixed with sugar.

To make about 3 pints [1 ½ liters]

4 lb.	ripe apples, peeled, cored and chopped (about 4 quarts [4 liters])	2 kg.
5 cups	sweet cider	1 ¼ liters
2¾ cups	brown sugar	675 ml.
1 tsp.	ground cloves	5 ml.
2 tsp.	ground cinnamon	10 ml.
½ tsp.	ground allspice	2 ml.

Put the cider into a pan and boil until it is reduced by half, about 30 minutes. Add the apples to the cider. Cook slowly until the fruit is tender, about 40 minutes; you will need to stir the mixture frequently. Then work the apple mixture through a sieve, and return the purée to the pan with the sugar and spices. Simmer for about 45 minutes, until the mixture thickens, stirring well. Pour the butter into jars and cover. Process *(pages 30-31).*

DAVID & ROSE MABEY
JAMS, PICKLES & CHUTNEYS

Apricot Butter

Marmelada od Kajsija

A high-quality apricot butter can be made only from fully ripe apricots. Hard, underripe apricots will not soften unless cooked with water.

To make about 3 cups [¾ liter]

2 lb.	large, very ripe apricots, halved and pitted (about 2 quarts [2 liters])	1 kg.
2 cups	sugar	½ liter
	rum or brandy	

Force the apricots through a sieve, discarding the peel; or purée them in a food grinder, including the peel. Place the fruit pulp in a bowl, sprinkle it with the sugar, cover, and leave the bowl in a cool place for two to three hours.

Transfer the fruit-and-sugar mixture to an enameled, tinned or stainless-steel pan and cook over high heat, stirring frequently, for 20 to 30 minutes, or until a wooden spatula leaves a path on the bottom of the pan. Spoon into hot jars. Cover the surface of the apricot butter with a paper disc dipped into a little rum or brandy. Seal the jars and store them in the refrigerator. The finished butter should be a beautiful golden yellow.

SPASENIJA-PATA MARKOVIČ (EDITOR)
VELIKI NARODNI KUVAR

Peach Butter

To make about 2 ½ pints [1 ¼ liters]

3 lb.	ripe peaches (8 to 10)	1 ½ kg.
½ cup	water	125 ml.
about 2½ cups	sugar	about 625 ml.
¼ tsp.	grated nutmeg	1 ml.
½ tsp.	ground cinnamon	2 ml.
½ tsp.	almond extract	2 ml.

Put the peaches in an enameled, tinned or stainless-steel pan, and pour in the water. Cover and bring to a boil, reduce the heat, and simmer for 20 minutes. Put aside to cool.

Lift the peaches out of the liquid, but do not discard the liquid. Slip off the peach skins, and remove the pits while putting chunks of the fruit into the blender. Purée the pulp and measure it: There should be about 5 cups [1¼ liters].

Return the purée to the pot with the liquid, and add ½ cup [125 ml.] of sugar for every cup [¼ liter] of purée. (I suggest reducing the total amount of sugar added by ½ cup, then tasting later for sweetness; different quality peaches require different amounts of sugar.) Add the nutmeg, cinnamon and almond extract, and return the uncovered pot to the heat. Cook slowly until the peach butter is very thick, about one and one half to two hours. Stir occasionally. To test its thickness, spoon some of the peach butter into a cold saucer; no ring of liquid should form around the edge. Taste for sweetness and add more sugar if needed, then simmer the peach butter for another 10 minutes. Put it into jars and process *(pages 30-31);* or cool the peach butter, spoon it into plastic containers, tightly cover and freeze.

CAROL CUTLER
THE SIX-MINUTE SOUFFLÉ AND OTHER CULINARY DELIGHTS

Beach-Plum Butter

To make about 3 pints [1 ½ liters]

4 lb.	beach plums, washed and pitted (about 2 quarts [2 liters])	2 kg.
	water	
2 cups	sugar	½ liter
1 tsp.	ground cinnamon	5 ml.
1 tsp.	ground cloves	5 ml.
½ tsp.	ground allspice	2 ml.
1	lemon, the juice strained and the peel finely chopped	1

In an enameled, tinned or stainless-steel pan, crush the plums to make the juices run. As the plums are likely to be dry, you will need to add some water before cooking them, but add as little as possible. Cook slowly until very soft—

about 15 minutes—then sieve the plums or put them through a food mill. Measure the pulp. Add ½ cup [125 ml.] of sugar to each cup [¼ liter] of pulp. Add the spices, lemon juice and peel. Cook the plum butter to the desired thickness, stirring often to prevent burning. Pour into hot jars and cover. Process *(pages 30-31)*.

GERTRUDE PARKE
GOING WILD IN THE KITCHEN

Plum Butter

Powidła ze Śliwek

This East European plum preserve should be made in August, when plums are at their sweetest.

To make about 7 pints [3 ½ liters]

9 lb.	plums, pitted (about 7 quarts [7 liters])	4 kg.
4 tbsp.	butter, melted	60 ml.
4 tbsp.	grated orange or lemon peel (optional)	60 ml.
1 ¾ cups	sugar	425 ml.

Coat the inside of a large saucepan with the melted butter. Then put in the plums and cook over low heat, stirring constantly with a large wooden spoon, for about 20 minutes, or until they are soft.

Put the plums through a strainer or food mill, clean the saucepan, and return the plum pulp to it. Stirring all the time, cook over low heat until the pulp is very thick and dark. Stir in the grated peel, if using, and the sugar. Put the butter into jars, cover and process *(pages 30-31)*.

E. KOLDER
KUCHNIA SLASKA

Plum Butter, Spiced

To make about 4 pints [2 liters]

6 lb.	plums, tart yellow or red, each slit 3 times (about 5 quarts [5 liters])	3 kg.
1 ½ cups	water	375 ml.
	sugar	
	ground cinnamon	
	ground allspice	
	grated nutmeg	
	ground cloves	

Place the slit plums and the water in an enameled, tinned or stainless-steel pan. Boil for 30 minutes, stirring frequently. Press through a sieve. Measure the fruit pulp. To every 1 cup [¼ liter] of pulp, allow 1 cup of sugar, ½ teaspoon [2 ml.] of cinnamon, ¼ teaspoon [1 ml.] of allspice, ⅛ teaspoon [½ ml.] of nutmeg and ⅛ teaspoon of cloves. Do not cook more than 4 cups [1 liter] of plum pulp at a time.

In a heavy saucepan, cook the sugar, spices and plum pulp, stirring constantly, for 20 minutes, or until the mixture is quite thick. Skim. Pour into jelly glasses or small jars. Cover and process *(pages 30-31)*. This spiced butter is fine with venison or any other game.

MARION FLEXNER
OUT OF KENTUCKY KITCHENS

Spiced Prune Butter

To make about 2 ½ pints [1 ¼ liters]

1 lb.	prunes, soaked in water overnight, drained and pitted	½ kg.
1 quart	water	1 liter
1	large orange	1
½ cup	seedless raisins, soaked in hot water for 15 minutes and drained	125 ml.
3 cups	sugar	¾ liter
½ tsp.	ground cloves	2 ml.
½ tsp.	ground cinnamon	2 ml.
¼ tsp.	ground ginger	1 ml.
½ cup	chopped pecans or walnuts	125 ml.

Combine the prunes with the water in a stainless-steel, tinned or enameled pan. Cut the peel from the orange; chop or cut it thin with scissors; add to the prunes. Slice the orange thin, cut the slices in half, seed the slices and add them to the prunes. Bring to a boil, and boil for 10 minutes. Add the raisins, sugar and spices. Stir over low heat until the sugar dissolves. Continue to cook and stir until thickened, about 10 minutes. Stir in the chopped nuts. Pour the butter into hot glass jars and cover. Process *(pages 30-31)*.

FLORENCE BROBECK
OLD-TIME PICKLING AND SPICING RECIPES

Pumpkin Jam

Confiture de Citrouille

To make about 3 pints [1 ½ liters]

2 lb.	pumpkin, quartered, seeded, peeled and diced (about 6 cups [1 ½ liters])	1 kg.
6	large apples, sliced with the peel and cores	6
7 ½ cups	water	1 ¾ liters
1 tsp.	vanilla extract, or the grated peel and strained juice of 1 lemon	5 ml.

Cook the apples in the water until they are pulpy; press them through a fine cloth and put the juice in a jug (there should be about 5 cups [1 ¼ liters]). Put the pumpkin pieces in a stainless-steel, enameled or tinned pan over very low heat, without water or sugar, and add the apple juice little by little. The jam should be flavored with either vanilla or lemon, according to taste, and cooked for at least five hours. Put the jam in jars and process *(pages 30-31)*.

X. MARCEL BOULESTIN
A SECOND HELPING OR MORE DISHES FOR ENGLISH HOMES

Apple and Tomato Cheese

This colorful cheese is excellent with duck or pork, and much liked as a preserve.

To make about 5 pints [2 ½ liters]

4 lb.	green cooking apples, chopped (about 4 quarts [4 liters])	2 kg.
2 lb.	green tomatoes, chopped (about 1 quart [1 liter])	1 kg.
2 cups	young spinach leaves, tied in a bunch (optional)	½ liter
	sugar	

Cook the apples and tomatoes in not more than 2 ½ cups [625 ml.] of water until quite soft and pulpy. A bunch of spinach leaves boiled with them ensures a good color.

Rub the apples and tomatoes through a sieve, or put them through a food mill, and add to them 1 ½ cups [375 ml.] of sugar for each 2 cups [½ liter] of pulp. Boil for 10 minutes, carefully stirring to prevent burning. Pour into hot jars. Cover and process *(pages 30-31)*.

THE DAILY TELEGRAPH
400 PRIZE RECIPES FOR PRACTICAL COOKERY

Damson Cheese

Damson cheese may be made with the pulp left in the jelly bag after jelly making. Strain the pulp to remove pits and skins; it will need no more cooking before the sugar is added and the mixture cooked as described in this recipe.

To make about 4 pints [2 liters]

4 lb.	damson plums (about 3 ½ quarts [3 ½ liters])	2 kg.
1 ¼ cups	water	300 ml.
	sugar	

Put the damson plums in a pan with the water, and simmer gently until quite soft, mashing the fruit occasionally. Purée the fruit through a sieve or the finest disk of a food mill. Measure the pulp, and measure 1 ¼ cups [300 ml.] of sugar for every 2 cups [½ liter] of pulp.

Return the pulp to the pan and cook very gently until thick, with no visible liquid. Warm the sugar slightly in a preheated 300° F. [150° C.] oven. Pour the warmed sugar into the pulp, stirring hard to dissolve it. Then increase the heat just a little and go on cooking and stirring until pressing the spoon down on top of the mixture leaves a mark. This produces a cheese firm enough to unmold and slice. For a very firm, almost candied consistency, go on cooking until a spoon drawn across the pan parts the mixture and the bottom shows very clearly. Spoon into oiled widemouthed jars and cover. Process *(pages 30-31)*.

JOCASTA INNES
THE COUNTRY KITCHEN

Guava Paste

Guayabate

For a drier consistency, place the cooked paste on wax paper on a wooden board, cover with cheesecloth, and set in the sun for two days, turning the paste from time to time to make sure all the surfaces are exposed to the sun.

To make about 2 pounds [1 kg.]

2 lb.	guavas, peeled	1 kg.
1 ½ cups	water	375 ml.
about 4 cups	sugar	about 1 liter

Cut the guavas in half and scoop out the seeds. Soak the seeds in 1 cup [¼ liter] of the water. Place the guavas in a saucepan with the remaining water, bring to a boil, reduce the heat to simmer, and cook the guavas until they are very soft. Take care that they do not scorch.

Strain the water from the seeds (it will be slightly mucilaginous), and add it to the cooked guavas. Discard the seeds.

Grind the guavas through the fine disk of a food grinder. Measure the pulp, and add an equal amount of sugar. Mix well; place in a large, heavy kettle over very low heat; and cook, stirring constantly with a wooden spoon, until the mixture is thick and a little jelly tested on a cube of ice can—when cold—be lifted off in one piece. Remove from the heat, and beat with a wooden spoon for 10 minutes, or until the mixture forms a heavy paste. Have ready a loaf pan, lined with wax paper. Turn the paste into the pan and set aside, in a cool place, for 24 hours. To store, turn the paste out of the pan and wrap it securely in foil.

ELISABETH LAMBERT ORTIZ
THE COMPLETE BOOK OF MEXICAN COOKING

Quince Cheese
Pasta de Membrillo

To make about 3 pints [1 ½ liters]

4	large quinces, wiped and shredded with a box grater	4
4 cups	sugar	1 liter
4 cups	water	1 liter

Prepare a syrup with the sugar and water, and boil together for five minutes. Add the quinces and cook over low heat, stirring frequently, for about two hours, or until the paste is translucent and thick. Put the paste into widemouthed jars, cover and process *(pages 30-31)*.

VICTORIA SERRA
TÍA VICTORIA'S SPANISH KITCHEN

Sour-Cherry Spoonsweet
Vyssino Glyko

To make about 2 ½ pints [1 ¼ liters]

2 to 2½ lb.	sour cherries, pitted (about 2 quarts [2 liters])	1 kg.
4 cups	sugar	1 liter
2 cups	water	½ liter
3 tbsp.	fresh lemon juice	45 ml.

Put the cherries in an enameled, tinned or stainless-steel pan in layers, sprinkling each layer with sugar. Add the water, and heat gently until the sugar has dissolved. Then increase the heat and boil for 25 minutes, stirring often and removing any scum that comes to the surface. Leave overnight in the syrup.

The next day, add the lemon juice, and boil until the syrup thickens and reaches a temperature of 235° F. [112° C.]. Spoon into jars, cover and process them *(pages 30-31)*.

CHRISSA PARADISSIS
THE BEST BOOK OF GREEK COOKERY

Watermelon-Rind Spoonsweet
Sladko ot Dini

Citric acid helps preserves to set without imparting a flavor of its own; it is available in pharmacies. To roast almonds, place them on a baking sheet in a 425° F. [220° C.] oven for five to 10 minutes, or until light brown. The watermelon flesh not used in this recipe may be added to fruit compotes or mixed preserves.

Traditionally eaten from a spoon and accompanied by coffee or water, spoonsweets can also be used as jam.

To make about 4 pints [2 liters]

8 lb.	watermelon, with a thick rind	3½ kg.
6 cups	sugar	1 ½ liters
3 cups	water	¾ liter
1 ½ tsp.	citric acid, dissolved in 1 ½ tbsp. [22 ml.] warm water	7 ml.
2 tbsp.	vanilla sugar, or ½ cup [125 ml.] blanched almonds, chopped and roasted	30 ml.

Cut the watermelon into wedges, then remove the flesh and seeds, and peel the thin green skin off the rind. Cut the thick rind into cubes, or use small cookie cutters to cut out various fancy shapes. Measure them. You should have about 8 cups [2 liters] of rind.

Bring a large pan of water to a boil, drop in the rind, and return to a boil. Drain the rind in a colander. Repeat this procedure twice more to soften the rind and eliminate any bitterness. Then spread the rind on a clean cloth, and leave it in the sun for 20 minutes, or until it turns ivory white.

Meanwhile, in a large enameled or stainless-steel pan, boil the sugar and the water together to a temperature of 220° F. [105° C.] as measured on a candy thermometer. Remove the pan from the heat, slide the rind into the syrup, and boil over moderately high heat until the syrup thickens and its temperature is 235° F. [112° C.]. Add the citric acid and vanilla sugar or almonds, and boil until the syrup returns to 235° F. Put into jars and process *(pages 30-31)*.

LIDIYA KURDZHIEVA (EDITOR)
DOMASHNO PRIGOTVYANE NA ZIMNINA

Vinegars and Sauces

Camp Vinegar

A recipe for walnut ketchup appears on page 124.

To make about 3 cups [¾ liter]

2½ cups	vinegar	625 ml.
¼ tsp.	cayenne pepper	1 ml.
3 tbsp.	soy sauce	45 ml.
¼ cup	walnut ketchup	50 ml.
1	small garlic clove, finely chopped	1
4	salt anchovies, filleted, rinsed, patted dry and chopped	4

Combine all the ingredients in a jar. Steep the vinegar for a month, shaking it every other day. Then strain the vinegar and pour it into bottles. Refrigerate.

ANNE COBBETT
THE ENGLISH HOUSEKEEPER

Nasturtium Vinegar

This vinegar is very useful as a flavoring for salads and sauces for lamb, etc.

To make 2½ quarts [2½ liters]

4 cups	freshly gathered nasturtium flowers	1 liter
2½ quarts	malt vinegar	2½ liters
4 or 5	whole cloves	4 or 5
8 to 10	black peppercorns	8 to 10
2 or 3	garlic cloves, halved	2 or 3
4 or 5	shallots, chopped	4 or 5

Small bottles should be used. Put all of the materials into the bottles, distributing the flavorings evenly among them, and cork and seal securely. Leave the vinegar in a dry place for a few months before using.

LIZZIE HERITAGE
CASSELL'S UNIVERSAL COOKERY BOOK

Pickle Vinegar

After the sun has extracted the essential oils from the flavoring spices, the vinegar may be strained and poured over parboiled or raw fruits or vegetables. If the fruits or vegetables are kept covered by the vinegar and stored in the refrigerator, they will keep for at least six months. As an alternative, the fruits or vegetables may be processed for 10 minutes in a boiling water bath.

Make this vinegar in May, and keep it in the sun all through the summer.

To make 10 quarts [10 liters]

8 quarts	vinegar	8 liters
2 cups	black mustard seeds	½ liter
½ cup	fresh ginger root	125 ml.
⅔ cup	whole allspice	150 ml.
3 tbsp.	whole cloves	45 ml.
¾ cup	black peppercorns	175 ml.
3 tbsp.	celery seed	45 ml.
3 lb.	brown sugar	1½ kg.
½ cup	freshly grated horseradish	125 ml.
2	garlic bulbs, separated into cloves and peeled	2
3	lemons, sliced	3

Combine all the ingredients in a large glass container. Close and set in the sun. Strain before using.

MARION CABELL TYREE
HOUSEKEEPING IN OLD VIRGINIA

Raspberry Vinegar

This vinegar, diluted with warm water, is an old remedy for sore throats. When there were colds about, it was often taken around as a nightcap to children in boarding schools. It is served in Yorkshire with Yorkshire pudding, as a sweet.

To make about 1 quart [1 liter]

2 lb.	raspberries, ripe and dry (about 4 pints [2 liters])	1 kg.
2½ cups	malt vinegar	625 ml.
about 4 cups	sugar	about 1 liter

Put the berries into a wide-necked jar and mash them well with a wooden spoon; then pour the cold vinegar onto them and leave, covered, for six days. Stir the mixture each day.

Now strain the raspberries through a jelly bag without pressing, and measure the liquid. Measure 2 cups [½ liter] of sugar to every 2½ cups [625 ml.] of liquid, and stir together over low heat until the sugar is dissolved. Boil gently for 10 minutes, removing any scum that rises. Leave until cold, then pour into bottles and cork firmly.

GERTRUDE MANN
BERRY COOKING

Queenie Williams' Pickling Syrup

This pickling syrup may simply be poured over raw or poached fruit to preserve it in the refrigerator for several months. This is enough for about 8 quarts [8 liters] of fruit.

This can be made up at the beginning of summer, put in a jar and kept in the refrigerator to use for pickling as needed. Or make it up whenever you wish.

To make about 2 quarts [2 liters]

4 cups	cider vinegar or white wine vinegar	1 liter
6 cups	sugar	1½ liters
1 cup	light corn syrup	¼ liter
24	whole cloves	24
1 tbsp.	coriander seeds	15 ml.
1	blade mace or 1 tsp. [5 ml.] ground mace	1
4 or 5	cinnamon sticks	4 or 5
2 tbsp.	mustard seeds or 2 tsp. [10 ml.] dry mustard	30 ml.
8	whole allspice	8

Tie the spices into a cheesecloth bag. Mix all of the other ingredients together in an enameled, tinned or stainless-steel pan. Add the bag of spices. Cook for 20 minutes from the time the syrup begins to boil. Cool. Remove the spice bag, pour the syrup into a jar, and keep in the refrigerator until needed, or use at once, if desired. If you prefer, the spices may be cooked in the syrup without a spice bag and the syrup may then be cooled and strained before pouring into jars.

MARION FLEXNER
OUT OF KENTUCKY KITCHENS

Violet Vinegar

Veilchen-Essig

This vinegar is used for sauces and stews. It can also be used to make a drink by adding 1 teaspoonful of violet vinegar to a glass of sugared water.

To make about 1 quart [1 liter]

½ cup	violet petals	125 ml.
1 quart	wine vinegar	1 liter

Place the violet petals in a glass bottle, pour in the vinegar, cork the bottle, and leave it in the sun or in a warm place for two to three weeks. Strain before using.

SOPHIE WILHELMINE SCHEIBLER
ALLGEMEINES DEUTSCHES KOCHBUCH FÜR ALLE STÄNDE

Apple Ketchup

This recipe is from *Library Ann's Cook Book*, compiled by the Minneapolis Public Library Staff Association.

To make about 2 pints [1 liter]

12	large, firm, tart apples, peeled, quartered and cored	12
1 cup	sugar	¼ liter
1 tsp.	ground white pepper	5 ml.
1 tsp.	ground cloves	5 ml.
1 tsp.	dry mustard	5 ml.
2	onions, finely chopped	2
2 cups	white vinegar	½ liter
2 tsp.	ground cinnamon	10 ml.
1 tbsp.	salt	15 ml.
½ cup	freshly grated horseradish	125 ml.

Place the apples in an enameled, tinned or stainless-steel pan, cover with water, and cook slowly, without a lid, until the apples are soft and the water has almost completely evaporated, about 30 minutes. Rub the apples through a sieve, or run them through a food mill. Add all of the other ingredients; heat to the boiling point, then reduce the heat to low and simmer slowly for one hour. Put in jars and process *(pages 30-31),* or place in plastic containers which have tight-fitting lids and keep refrigerated.

LOUIS SZATHMARY
AMERICAN GASTRONOMY

Blackberry Ketchup

To make about 5 pints [2 ½ liters]

4 lb.	ripe blackberries (about 3½ quarts [3½ liters])	2 kg.
2 lb.	brown sugar	1 kg.
2 cups	vinegar	½ liter
2 tsp.	ground cloves	10 ml.
2 tsp.	ground cinnamon	10 ml.
1 tsp.	ground allspice	5 ml.

Cook the blackberries slowly for two hours with the sugar, vinegar, cloves, cinnamon and allspice. When all is quite soft, put into jars and cover. Process *(pages 30-31)*.

THE DAILY TELEGRAPH
400 PRIZE RECIPES FOR PRACTICAL COOKERY

Lemon Ketchup

A few sliced ripe tomatoes, packed into the jars with the lemons, are considered by some to improve both the color and the flavor of the ketchup. A few drops of this ketchup will give savor to hosts of dishes. Should it be wanted for immediate use, increase the quantities of spices, and boil them all with the vinegar.

To make about 2 ½ pints [1 ¼ liters]

6	lemons, peeled	6
⅓ cup	salt	75 ml.
3 tbsp.	shallots, finely chopped	45 ml.
1	garlic clove, finely chopped	1
3½ tsp.	ground mace	17 ml.
1 tsp.	whole cloves, crushed	5 ml.
2 tbsp.	ground ginger	30 ml.
1 tsp.	cayenne pepper	5 ml.
⅔ cup	freshly grated horseradish	150 ml.
3½ cups	white vinegar	825 ml.

Cut off a piece from both ends of each lemon, or make very deep incisions in the lemons, and rub in the salt. The surplus salt should be rubbed over the outside of the lemons. Put them in a jar with the shallots, garlic and spices, reserving a little of the mace and ginger. Add the horseradish.

Boil the vinegar for five minutes with the reserved mace and ginger, and pour this over the lemons. Cover lightly and, when cold, close the jar tightly. Refrigerate. Strain after six months—or better still, after 12 months. The strained ketchup should be put into small bottles with new corks.

LIZZIE HERITAGE
CASSELL'S UNIVERSAL COOKERY BOOK

Old-fashioned Tomato Ketchup

Those who like the flavor of onions may add about half a dozen medium-sized ones, peeled and sliced, 15 minutes before the vinegar and spices are put in.

To make about 22 pints [11 liters]

33 lb.	tomatoes, sliced (about 15 quarts [15 liters])	15 kg.
½ cup	salt	125 ml.
¾ cup	ground black pepper	175 ml.
⅓ cup	ground cinnamon	75 ml.
2 tbsp.	ground cloves	30 ml.
1 tbsp.	cayenne pepper	15 ml.
⅔ cup	sugar	150 ml.
4 quarts	vinegar	4 liters

Stew the tomatoes in their own liquor until soft, and rub through a sieve fine enough to retain the seeds. Boil the pulp and juice down to the consistency of apple butter (very thick), stirring steadily to prevent burning. Mix the spices and sugar with the vinegar, then add to the tomatoes. Boil up twice, then bottle. Process *(pages 30-31)*.

THE BUCKEYE COOKBOOK

Wild-Plum Ketchup

This recipe is taken from the Presbyterian Cook Book, compiled by the ladies of the First Presbyterian Church, Dayton, Ohio. Wild plums are a native American fruit, found in most eastern and central states. They ripen in August and September. Tart cultivated plums may be substituted.

To make 8 pints [4 liters]

10 lb.	plums, stemmed (about 7½ quarts [7½ liters])	4½ kg.
10 cups	sugar	2½ liters
1 quart	white vinegar	1 liter
1 tsp.	ground cinnamon	5 ml.
½ tsp.	grated nutmeg	2 ml.
½ tsp.	ground cloves	2 ml.

Wash your bottles; keep them warm until ready for use. Add the sugar to the plums, and place them in an enameled, tinned or stainless-steel pan over medium heat. Cover. After 30 minutes, increase the heat and stir the plum-sugar mixture. Break some of the plums with the wooden spoon.

Continue cooking, stirring the mixture occasionally so that it does not stick to the pan. In approximately 15 minutes, all of the plums will be cooked through and mashed. Remove from the heat and cool.

Mash the mixture with a wooden spoon. Put the mashed

pulp through a fine sieve, and continue mashing until all of the liquid and some of the pulp go through. Discard the pits and the skins. There should be about 4 quarts [4 liters] of liquid. Boil the vinegar together with the spices for a few minutes. Add the vinegar-spice mixture to the plum liquid, then boil vigorously, stirring constantly, for 15 minutes, or until the liquid has reduced to 4 quarts again. Pour into bottles or jars. Process *(pages 30-31)*.

LOUIS SZATHMARY
AMERICAN GASTRONOMY

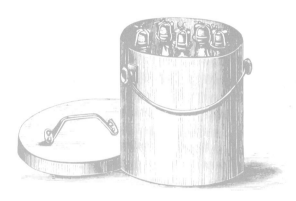

To Make English Ketchup

This recipe for ketchup is taken from a book published anonymously in 1758, but generally attributed to the English cookery writer Hannah Glasse. The original version calls for "strong stale mummy Beer," meaning beer that has almost turned to vinegar. Today, ordinary vinegar should be used.

To make about 2 ½ pints [1 ¼ liters]

6 lb.	mushrooms, broken into pieces (about 7 ½ quarts [7 ½ liters])	3 kg.
1 cup	salt	¼ liter
2½ cups	vinegar	625 ml.
20	salt anchovies	20
1 tsp.	ground mace	5 ml.
1 tsp.	ground cloves	5 ml.
1 tbsp.	pepper	15 ml.
1	slice fresh ginger	1
1½ cups	shallots, sliced	375 ml.

Mix the mushrooms with the salt in a bowl, cover, and let them stand for nine days, stirring them once or twice a day. Put them into a stoneware crock, cover tightly, and set the crock in a pan of water. Cook over low heat for three hours. Strain the mushrooms through a sieve into a pot, and add to the juice the remaining ingredients—the proportions given are for 5 cups [1¼ liters] of mushroom liquid. Keeping the pot covered, boil together over low heat until the liquid is reduced by half, then strain it through a jelly bag. Bottle the ketchup and cover. Process *(pages 30-31)*.

THE ART OF COOKERY, MADE PLAIN AND EASY

Prince of Wales Ketchup

To prepare elderberry vinegar, strip about 2 cups [½ liter] of ripe berries from their stems. Place in an ovenproof dish. Cover with vinegar. Cook in a 300° F. [150° C.] oven for one and one half hours, or until the berries burst and become very juicy. Let the mixture stand overnight, then strain it through a jelly bag.

To make 1 pint [½ liter]

2½ cups	elderberry vinegar	625 ml.
5	salt anchovies, filleted, rinsed, patted dry and chopped	5
3 tbsp.	shallots, thinly sliced	45 ml.
1 tsp.	whole cloves	5 ml.
1 tsp.	ground mace	5 ml.
1 tsp.	ground ginger	5 ml.
1 tsp.	grated nutmeg	5 ml.

Place all the ingredients in an enameled, tinned or stainless-steel pan. Bring to a boil. Simmer gently until the anchovies are broken up, about 20 minutes. Let cool; strain and bottle. Cork tightly. Refrigerate. Leave for two weeks before opening. The ketchup will keep for about six months.

ELIZABETH CRAIG
COURT FAVOURITES

Oyster Ketchup

Mussels may be used instead of oysters. A pounded anchovy or two may be added to give flavor.

To make 2 ½ pints [1 ¼ liters]

16	live oysters, shucked, with the liquor reserved	16
2½ cups	dry sherry	625 ml.
2 tbsp.	salt	30 ml.
½ tsp.	ground mace	2 ml.
¼ tsp.	cayenne pepper	1 ml.
1 tbsp.	brandy (optional)	15 ml.

Pound the oysters in a mortar, and add the oyster liquor and the sherry. Bring the mixture to a boil, then add the salt, mace and cayenne pepper. Boil up again, skim, then strain through a sieve. Stir in the brandy, if used. Put in jars and cover. The ketchup will keep for six weeks if refrigerated.

ANNE COBBETT
THE ENGLISH HOUSEKEEPER

Lobster Ketchup

Add 4 to 5 tablespoons [60 to 75 ml.] of this ketchup to 1 cup [¼ liter] of melted butter as a sauce for fish dishes. For chili vinegar, use 10 dried hot chilies. Shallot vinegar is made by steeping eight chopped shallots in 2½ cups [625 ml.] of wine vinegar for three weeks; shallot wine by steeping the shallots in dry white wine.

To make about 3 ½ pints [1 ¾ liters]

one 3 lb.	female lobster, boiled in water to cover for 10 minutes, drained, body shell cut lengthwise in half, claws cracked with a mallet or nutcracker	one 1 ½ kg.
6	salt anchovies, filleted, rinsed and patted dry	6
1 ¼ cups	sherry or Madeira	300 ml.
1 tsp.	cayenne pepper	5 ml.
⅔ cup	chili or shallot vinegar	150 ml.
3 ½ cups	shallot wine	875 ml.
	black peppercorns	

Pick out all the lobster meat, and pound the coral (roe) with the anchovy fillets in a mortar. When completely bruised, add the lobster meat; pound and moisten it with the remaining ingredients. Mix well, and put the ketchup into wide-mouthed jars. Put 1 teaspoon [5 ml.] of peppercorns into each jar. Close tightly. Refrigerated, the ketchup will keep for up to two months.

ANNE COBBETT
THE ENGLISH HOUSEKEEPER

Walnut Ketchup

To make about 3 pints [1 ½ liters]

24	green walnuts	24
¾ cup	salt	175 ml.
5 cups	water	1 ¼ liters
2 quarts	vinegar	2 liters
2 tsp.	ground cloves	10 ml.
2 tsp.	ground mace	10 ml.
12	garlic cloves	12

Put the walnuts and salt into the water and leave them for nine days. Remove the walnuts from the brine and pound them in a mortar. Combine the walnuts and vinegar and leave them for a week, stirring every day.

Strain the mixture through a muslin bag, squeezing to extract all the liquor. To this juice add the cloves, mace and

garlic; boil it for 15 to 20 minutes, strain, and then bottle it. Cover and process *(pages 30-31).*

MISS TYSON
THE QUEEN OF THE KITCHEN

Thick Provençal Tomato Sauce

Coulis de Tomates à la Provençale

The juice squeezed out of the tomatoes before they are cooked can be saved for drinking. Since there are no preservatives in the tomato juice, it will ferment in a few days, even in the refrigerator. To avoid that, pour the juice into single-portion-sized plastic glasses, cover and freeze. Even in the dead of winter this juice will bring back bright memories of summer sunshine.

To make about 5 pints [2 ½ liters]

6 lb.	tomatoes	2 ¾ kg.
⅓ cup	olive oil	75 ml.
1 cup	finely chopped onion	¼ liter
⅓ cup	flour	75 ml.
½ tsp.	sugar	2 ml.
5	garlic cloves, mashed	5
3	1-inch [2½-cm.] pieces orange peel, all pith removed	3
1	herb bouquet, made of 6 parsley sprigs tied around 2 bay leaves	1
½ tsp.	fennel seeds	2 ml.
½ tsp.	dried basil	2 ml.
¼ tsp.	ground coriander	1 ml.
1 tsp.	salt	5 ml.
½ tsp.	pepper	2 ml.
½ tsp.	celery salt	2 ml.
	ground saffron	
2 to 4 tbsp.	tomato paste	30 to 60 ml.

Plunge the tomatoes into boiling water for a few seconds and remove the skins; then cut the tomatoes in half and gently squeeze out the seeds and juice. Chop the remaining tomato pulp coarse. There should be 7 to 8 cups [1¾ to 2 liters].

Heat the oil in a heavy enameled pot, add the onions, and cook slowly for about 15 minutes with the lid on. Do not let the onions brown; they should just be transparent.

Remove the lid and stir in the flour, and cook the mixture for two or three minutes, still not allowing it to brown. Add the tomato pulp and all of the seasonings and herbs, cover the pot, and simmer slowly for 15 minutes.

Add 2 tablespoons [30 ml.] of the tomato paste, and simmer the sauce very slowly, partially covered, for one to one and one half hours. Stir occasionally, making certain to scrape the bottom of the pot so that the mixture does not stick or scorch as it thickens. As you stir, press the tomato pulp against the sides of the pot to make the sauce smoother. If there is any risk of scorching, add a little of the tomato juice. Cook until the sauce is very thick and will stand up in a mass on a metal spoon.

Remove the herb bouquet and the orange peel. If the color seems a little pale, or if you would like to intensify the tomato flavor, add the remaining 2 tablespoons of tomato paste and cook 10 minutes longer. Taste for seasoning and add more salt and pepper if necessary.

To freeze, first cool the *coulis* to room temperature. Then spoon it into 1- or 2-cup [¼- or ½-liter] plastic containers, allowing ½ inch [1 cm.] headspace, cover the containers with lids, and freeze.

To can, spoon the *coulis* while still hot into clean hot canning jars, allowing ½ inch headspace, close, and process for 45 minutes in a boiling water bath (the time is the same for either pints or half pints).

JAMES BEARD, MILTON GLASER, BURTON WOLF (EDITORS)
THE GARDEN-TO-TABLE COOKBOOK

Tomato Purée

Conserva di Pomidoro

To make about 8 cups [2 liters]

9 lb.	tomatoes, halved (about 4 quarts [4 liters])	4 kg.
2 cups	finely chopped celery	½ liter
2 cups	finely shredded carrots	½ liter
⅓ cup	salt	75 ml.
1 cup	olive oil	¼ liter

Place the tomatoes, celery and carrots in a saucepan, cover, and cook over medium heat for about 30 minutes, or until all the vegetables are soft. Purée through a sieve or food mill, and put the purée in a jelly bag to drain for 10 to 12 hours.

Put the drained, thick purée in a heavy pan with the salt and the oil. Stirring frequently, simmer for about 30 minutes, or until the purée drops very slowly from a spoon. Pack the purée into jars, cover and process *(pages 30-31)*.

EMMANUELE ROSSI (EDITOR)
LA VERA CUCINIERA GENOVESE

Honey and Apple Conserve

This conserve is good with cold meat of any kind.

To make about 3 pints [1 ½ liters]

1¼ lb.	honey	600 g.
2 lb.	tart apples, peeled, cored and quartered (about 2 quarts [2 liters])	1 kg.
2½ cups	cider vinegar	625 ml.
2 tsp.	ground cloves	10 ml.
2 tsp.	ground cinnamon	10 ml.

Heat together the honey, vinegar, cloves and cinnamon; then add the apples. Stirring frequently, boil for 45 minutes, until the mixture is thick and smooth. Pour into warm jars and cover at once. Process *(pages 30-31)*.

GERTRUDE MANN
THE APPLE BOOK

Spiced Cantaloupe

To make 1 quart [1 liter]

2 lb.	firm, ripe cantaloupe	1 kg.
⅔ cup	sugar	150 ml.
⅓ cup	water	75 ml.
3 tbsp.	white vinegar	45 ml.
2-inch	stick cinnamon	5-cm.
4	whole cloves	4

Cut the cantaloupe into quarters and, with a spoon, scoop out the seeds and stringy pulp. With a small, sharp knife, remove the skin and the inner rind. Then cut the meat into 2-inch [5-cm.] pieces, and pack the cantaloupe pieces into a 1-quart [1-liter] canning jar.

Combine the sugar, water, vinegar, cinnamon and cloves in an enameled, tinned or stainless-steel pan, and bring to a boil over high heat, stirring until the sugar dissolves. Cook briskly, uncovered, for five minutes.

With tongs, remove the stick cinnamon and tuck it down the side of the cantaloupe-filled jar. Then ladle the hot liquid over the cantaloupe a few tablespoonfuls at a time, allowing the liquid to flow to the bottom of the jar before adding more. Fill the jar to within ½ inch [1 cm.] of the top. Cover, and process the jar for 12 minutes in a boiling water bath.

FOODS OF THE WORLD/AMERICAN COOKING: THE EASTERN HEARTLAND

Bitter Lemon Pickle

To make about 4 pints [2 liters]

8	lemons, each cut lengthwise into 8 wedges and seeded	8
2 tbsp.	salt	30 ml.
4 cups	seedless white raisins, chopped	1 liter
1 tbsp.	chopped fresh ginger	15 ml.
1 tsp.	cayenne pepper	5 ml.
4	garlic cloves, crushed to a paste	4
1¼ cups	cider vinegar	300 ml.
4 cups	light brown sugar	1 liter
2 tsp.	freshly grated horseradish	10 ml.

Sprinkle the lemon wedges with the salt, and refrigerate for at least 36 hours, stirring occasionally. Drain the lemons and reserve the liquid. Put the lemons through a food grinder, using the coarse disk, and combine with the reserved liquid and all of the other ingredients in a saucepan. Simmer over low heat until the mixture is thick, one to one and one half hours. Put into jars, cover and process *(pages 30-31)*.

PATRICIA HOLDEN WHITE
FOOD AS PRESENTS

Sweet-and-Sour Pears

Süss-saure Birnen

To make about 3 quarts [3 liters]

5 lb.	pears, peeled, halved lengthwise and cored (about 3 quarts [3 liters])	2½ kg.
1 quart	wine vinegar or cider vinegar	1 liter
4½ cups	sugar	1⅛ liters
5	whole cloves	5
½	cinnamon stick	½
1	lemon, the peel only, cut in a spiral	1
1 cup	water	¼ liter

To keep the pears from turning brown as you prepare them, place them in water mixed with a little vinegar.

Combine the 1 quart [1 liter] of vinegar, the sugar, cloves, cinnamon and lemon peel with the 1 cup [¼ liter] of water. Bring this syrup to a boil. Drain and add the pears, and cook over low heat for about 15 minutes, or until the pears are tender but still firm. Remove the pears from the syrup and pack them into jars.

Cook the syrup for another 15 minutes, then pour it over the pears in the jars. After three days, pour off the syrup, boil it for three minutes, and pour it back over the pears. Cover the jars and process them *(pages 30-31)*.

HANS KARL ADAM
DAS KOCHBUCH AUS SCHWABEN

Sweet Pickled Peaches, Plums or Pears

If you use peaches for this, wash them and rub off the fuzz. For plums, prick the skins. Be sure any pears are firm and not overly ripe. Wash them and remove the blossom ends.

To make about 7 pints [3 ½ liters]

8 lb.	peaches, plums or pears (about 3½ quarts [3½ liters])	3½ kg.
⅓ cup	whole cloves	75 ml.
4-inch	stick cinnamon, broken	10-cm.
1 tbsp.	chopped fresh ginger	15 ml.
4 lb.	brown sugar	2 kg.
4 cups	cider vinegar	1 liter

Tie the cloves, cinnamon and ginger in a piece of muslin or cheesecloth to make a spice bag. Boil the sugar in the vinegar with the spice bag for about 10 minutes, or until you have a thick syrup. Pour the syrup over the fruit and let stand overnight.

Drain off the syrup and boil it for 10 minutes. Add the fruit to it and cook, uncovered, until just tender. The time will depend on the fruit used. Put in hot jars with glass or metal lids. Process in a boiling water bath for 15 minutes.

JANE MOSS SNOW
A FAMILY HARVEST

Pickled Plums

Quetsches Confites au Vinaigre

This is a Belgian pickle to serve with meats. It is made from *quetsches*, small purple plums.

To make about 6 pints [3 liters]

4 lb.	slightly underripe plums, each pricked in several places with a needle (about 3 quarts [3 liters])	2 kg.
8 cups	sugar	2 liters
1 cup	water	¼ liter
2½ cups	vinegar	625 ml.
5	cinnamon sticks, broken into small pieces	5
⅓ cup	whole cloves	75 ml.

Bring the sugar and water to a boil over high heat, and cook for about 10 minutes to make a clear syrup. Add the vinegar, cinnamon and cloves, and boil for five minutes more.

Add the plums and bring the mixture to a boil over medium heat; to avoid breaking the fruit do not boil it hard. Skim.

Remove the fruit from the syrup with a skimmer, then boil the syrup over high heat for five minutes. Remove from the heat, return the plums to the syrup and allow the mixture to cool. Refrigerate for 24 hours.

The next day, bring the mixture to a boil, remove the plums, boil the syrup for five minutes, return the plums to the pan and let the mixture cool. Let stand another 24 hours. Put the plums into jars, cover and process *(pages 30-31)*. Store for at least six weeks before using.

E. AURICOSTE DE LAZARQUE
CUISINE MESSINE

Watermelon Pickle

To make about 2 pints [1 liter]

2 lb.	watermelon rind, green skin and pink flesh pared off, rind cut into thin strips or rounds (about 7 cups [1¾ liters])	1 kg.
3 tbsp.	salt	45 ml.
1 quart	boiling water	1 liter
2 cups	white or cider vinegar	½ liter
6 cups	sugar	1½ liters
1 tsp.	ground cinnamon	5 ml.
1 tsp.	ground cloves	5 ml.
2 tbsp.	whole cloves	30 ml.
2 tbsp.	whole allspice	30 ml.
five 3-inch	cinnamon sticks, broken	five 8-cm.

Mix 1 quart [1 liter] of cold water with the salt, and pour over the rind. Cover and let stand overnight. Drain; cover the rind with 1 quart of fresh water, and cook for eight to 10 minutes. Drain the rind again.

In a stainless-steel, tinned or enameled pan, combine the quart of boiling water, the vinegar, sugar and ground spices. Tie the whole spices in a small cheesecloth bag and add to the pan. Bring to a boil and stir until the sugar dissolves. Boil for five minutes. Add the drained rind. Boil gently until the rind is transparent, about 45 minutes. Remove the spice bag. Spoon the fruit into hot jars; cover at once. Process *(pages 30-31)*. Let the jars stand for about four weeks before using.

FLORENCE BROBECK
OLD TIME PICKLING AND SPICING RECIPES

Artichokes Preserved in Oil

Carciofini sott'Olio

To make about 6 pints [3 liters]

4 lb.	small artichokes (4 quarts [4 liters])	2 kg.
4 cups	white wine vinegar	1 liter
1 tsp.	salt	5 ml.
½ tsp.	black peppercorns	2 ml.
4	whole cloves	4
2½ cups	olive oil	625 ml.

Put the vinegar in an enameled, tinned or stainless-steel pan with the salt, peppercorns and cloves. Remove the tough outside leaves from each artichoke until the pale inner leaves are exposed, cut off the tips of the remaining leaves, and trim the base to a pointed shape. As each artichoke is ready, place it in the vinegar.

Boil the artichokes for about 10 minutes or until tender, drain and cool. Place the artichokes in glass jars, scattering the peppercorns and cloves among them. Cover the artichokes with the oil, put on lids and process *(pages 30-31)*.

GIUSEPPE OBEROSLER (EDITOR)
IL TESORETTO DELLA CUCINA ITALIANA

Crisp Dilled Beans

To make 4 pints [2 liters]

2 lb.	tender, mature green or yellow beans, trimmed but left whole (about 2 quarts [2 liters])	1 kg.
1 tsp.	cayenne pepper	5 ml.
4	garlic cloves	4
4	large heads fresh dill	4
2 cups	water	½ liter
¼ cup	coarse salt	50 ml.
2 cups	cider vinegar	½ liter

Pack the beans uniformly into four 1-pint [½-liter] jars, stem ends down. To each jar, add ¼ teaspoon [1 ml.] of cayenne pepper, one garlic clove and one head of dill.

Combine the water, salt and vinegar in an enameled, tinned or stainless-steel pan and bring to a rolling boil. Pour the liquid over the beans, filling the jars to within ½ inch [1 cm.] of the tops. Wipe the rims; adjust the lids. Process the jars in a boiling water bath for 10 minutes. Start to count the processing time when the water returns to a boil. Remove the jars and complete the seals. These pickles are ready to serve immediately.

NELL B. NICHOLS AND KATHRYN LARSON (EDITORS)
FARM JOURNAL'S FREEZING & CANNING COOKBOOK

Pickled Beets

You can pickle either small baby beets, which are the thinnings from the crop, and also the very large roots at the end of the season. In both cases it is important not to bruise the beets. Twist the leaves off by hand, leaving a little stem, and wash but do not peel the beets. This recipe will give you a pickle with a fresh, fiery taste. The pickle is ready for use soon after it is made.

To make about 1 quart [1 liter]

2 lb.	beets	1 kg.
	salt	
about 2½ cups	white vinegar	about 625 ml.
½ tsp.	cayenne pepper	2 ml.
1 tsp.	ground ginger	5 ml.
3 or 4	black peppercorns	3 or 4

Boil the beets in a saucepan of salted water until they are tender, about 15 minutes for baby beets and up to one hour for large ones. Meanwhile, bring the vinegar and spices to a boil in an enameled, tinned or stainless-steel pan.

When the beets are cooked, let them cool a little before peeling them. Put small ones whole into a widemouthed glass jar and pour the spiced vinegar over them so that they are completely covered. When you are pickling large, old beets, you will need to cut them into slices before pickling. Close the jar and process *(pages 30-31)*.

DAVID & ROSE MABEY
JAMS, PICKLES & CHUTNEYS

Dill Pickles
Suolakurkut

To make 2 quarts [2 liters]

7 to 10	pickling cucumbers, washed	7 to 10
4 cups	black currant, oak or cherry leaves, rinsed	1 liter
1 tbsp.	grated fresh horseradish	15 ml.
1 bunch	dill stems, leaves removed	1 bunch
5 to 7	heads fresh dill or 1 tsp. [5 ml.] dill seeds	5 to 7
1 quart	water	1 liter
½ cup	white vinegar	125 ml.
¼ cup	coarse salt	50 ml.
1 tsp.	white peppercorns	5 ml.
1	garlic clove, peeled and crushed	1

Wrap each cucumber in black currant or other leaves. Fit the cucumbers into a 2-quart [2-liter] jar. Between them, put the horseradish and the dill stems and heads, if used. Combine the water, vinegar, salt, peppercorns, garlic and dill seeds, if used, in a saucepan. Bring to a boil, reduce the heat and simmer for a couple of minutes, then cool. Pour the cold marinade over the cucumbers. They should be totally covered. Refrigerate for two to three weeks before serving; they will keep for four to five weeks longer in the refrigerator. If processed *(pages 30-31)* they can be kept for up to a year.

ULLA KÄKÖNEN
NATURAL COOKING THE FINNISH WAY

Brined Dill Pickles

For a strong garlic flavor, add 10 to 20 garlic cloves to the pickling brine. For a mild garlic flavor, add a garlic clove to each jar of pickles before processing it.

To make about 10 quarts [10 liters]

20 lb.	cucumbers, each 3 to 6 inches [8 to 15 cm.] long, scrubbed, blossom ends removed (about ½ bushel)	9 kg.
¾ cup	pickling spices	175 ml.
2 or 3	bunches fresh or dried dill	2 or 3
2½ cups	vinegar	625 ml.
1¾ cups	coarse salt	425 ml.
10 quarts	water	10 liters

Place half of the pickling spices and a layer of the dill in a 5-gallon [20-liter] crock. Fill the crock with cucumbers to within 3 to 4 inches [8 to 10 cm.] of the top. Place the remaining dill and pickling spices over the top of the cucumbers. Thoroughly mix the vinegar, salt and water, and pour this pickling brine over the cucumbers.

Cover the cucumbers with a heavy china or glass plate or lid that fits inside the crock. Use a weight to hold the plate down and keep the cucumbers under the brine. A glass jar filled with water makes a good weight. Cover loosely with a clean cloth. Keep the pickles at room temperature, and remove the scum daily after it begins to form; scum may start forming in three to five days. Do not stir the pickles, but be sure that they are completely covered with brine. If necessary, make additional brine, using the original proportions of vinegar and salt to water.

In about three weeks, the cucumbers will become olive-green and should have a desirable flavor. Any white spots inside the cucumbers will disappear in processing.

The original brine is usually cloudy as a result of yeast that develops during the fermentation period. If this cloudiness is objectionable, fresh brine may be used to cover the pickles when packing them into jars; in making fresh brine, use ½ cup [125 ml.] of salt and 4 cups [1 liter] of vinegar to 4 quarts [4 liters] of water. However, the fermentation brine is generally preferred for its added flavor; strain it before heating it to boiling.

Pack the pickles, along with some of the dill from the crock, into clean, hot quart [1-liter] jars. Avoid too tight a pack. Boil brine, and pour it into the jar to within ½ inch [1 cm.] of the tops. Adjust the jar lids.

Process the pickles in a boiling water bath for 15 minutes; start to count the processing time as soon as the hot jars are placed into the actively boiling water.

Remove the jars, and complete the seals, if necessary. Set the jars upright, several inches apart, on a wire rack to cool.

U.S. DEPARTMENT OF AGRICULTURE
COMPLETE GUIDE TO HOME CANNING, PRESERVING, AND FREEZING

———————◆———————

Ada Gail's Authentic Jewish Pickled Cucumber

The techniques of making pickles are shown on pages 86-87.

In the sugarless brine used here, following the same procedure, you can also pickle green beans, shallots, green tomatoes, baby ears of corn and small peppers. Alum is available at drugstores.

To make about 6 quarts [6 liters]

6 lb.	pickling cucumbers (about 5 quarts [5 liters]), each about 3 inches [8 cm.] long, scrubbed	3 kg.
18	sprigs fresh dill	18
18	garlic cloves, peeled	18
6	fresh hot chilies, stemmed and seeded	6
36	black peppercorns	36
1 tbsp.	pickling spices	15 ml.
1½ tsp.	alum	7 ml.
8½ quarts	water	8½ liters
1½ cups	coarse salt	375 ml.
½ cup	cider vinegar	125 ml.
6	fresh grapevine leaves (optional)	6

In each of six 1-quart [1-liter] canning jars, place three sprigs of dill, three garlic cloves, one hot chili, six peppercorns, a pinch of pickling spices and, to keep the pickles crisp, a pinch of alum. Pack the cucumbers into the jars.

Prepare the brine by boiling the water with the salt and vinegar. Pour the boiling brine over the cucumbers to immerse them. Reserve the remaining brine. Place a grapevine leaf, if available, over the contents of each jar and put on lids. Set the jars aside at room temperature.

Wait for the brine to ferment (bubbles should appear after two or three days), bring the reserved brine to a boil and use it to fill the jars—to within ½ inch [1 cm.] of the top. Close the jars, set them aside at room temperature, and wait for the brine to clear, indicating that fermentation has ceased. Refrigerate the pickles; you can eat them after four or five days, and they will keep for two months. Processed *(pages 30-31)*, the pickles will keep for at least a year.

PAUL LEVY
HABITAT COOK'S DIARY 1980

Seven-Day Sweet Pickles

To make about 7 pints [3½ liters]

7 lb.	pickling cucumbers, about 3 inches [8 cm.] long (about 5 quarts [5 liters])	3½ kg.
1 quart	cider vinegar	1 liter
8 cups	sugar	2 liters
2 tbsp.	salt	30 ml.
2 tbsp.	pickling spices	30 ml.

Wash the cucumbers, put them into a large container and cover them with boiling water. Let stand for 24 hours. Drain. Repeat this process each day for four days, using fresh boiling water each time.

On the fifth day, slice the cucumbers into ¼-inch [6-mm.] rings. Combine the vinegar, sugar, salt and pickling spices in a stainless-steel, tinned or enameled pan. Bring to a boil, and pour this syrup over the sliced cucumbers. Let stand for 24 hours. On the sixth day, drain off the syrup, bring it to a boil and pour it over the cucumbers.

On the seventh day, drain off and boil the syrup once more. Add the cucumbers and bring the syrup back to the boiling point. Pack the pickles and the syrup into hot jars, and cover. Process *(pages 30-31)*. Store the pickles for two weeks before opening the jars.

BETTY GROFF AND JOSÉ WILSON
GOOD EARTH & COUNTRY COOKING

———————◆———————

West Indian Pickles

To make about 5 quarts [5 liters]

5 lb.	small pickling cucumbers (about 4½ quarts [4½ liters]), scrubbed	2½ kg.
½ cup	salt	125 ml.
2	onions, sliced	2
1	sweet red pepper, halved, seeded, deribbed and sliced	1
4	garlic cloves, peeled and sliced	4
3	small fresh hot chilies	3
1 tbsp.	ground cardamom	15 ml.
3 tbsp.	black peppercorns	45 ml.
3 quarts	cider vinegar	3 liters

Cover the cucumbers with water, add the salt and let the cucumbers soak overnight. The next day, drain and rinse the cucumbers. Place the cucumbers in jars with the onions, sweet pepper slices, garlic and whole chilies. Combine the cardamom and peppercorns with the vinegar and bring to a boil. Pour the vinegar mixture into the jars to within ½ inch [1 cm.] of the tops. Process *(pages 30-31)*.

CONNIE AND ARNOLD KROCHMAL
CARIBBEAN COOKING

Cornichons by the Cold Method

Though special varieties of cucumbers are grown in France for this pickle, it is possible to achieve satisfactory results using the smallest pickling cucumbers to be found on the market. If you grow your own, pick them while still immature and 2 to 2½ inches [5 to 6 cm.] in length.

To make about 10 pints [5 liters]

6 to 7 lb.	small, freshly picked cucumbers, rubbed with a towel to remove spikes (about 5 quarts [5 liters])	3 kg.
2 cups	salt	½ liter
	garlic cloves	
	small boiling onions	
	thyme and tarragon sprigs	
	bay leaves	
	black and white peppercorns	
	whole cloves	
2 quarts	white wine vinegar	2 liters

Place the cucumbers in a stainless-steel or earthenware dish and sprinkle with the salt. Leave for one day to soak, but turn the cucumbers frequently.

At the end of this time, the cucumbers will be totally limp and flexible. Wipe each one to remove salt. Rinse preserving jars with boiling water and turn upside down to drain. Place the cucumbers in the jars, and add to each jar: two whole, peeled garlic cloves, several boiling onions, a sprig of thyme and/or tarragon, a bay leaf, and half a teaspoon [2 ml.] each of whole peppercorns and cloves. Cover with the vinegar, close the jars, and store for five weeks before tasting. This method will produce strong, sharply acid pickles, which will keep for up to six months.

JUDITH OLNEY
SUMMER FOOD

Pickled Cucumbers

Cornichons à la Française

Surely everyone interested in the French cuisine has, at one time or another, bought a jar of these excellent tiny pickled cucumbers. Mademoiselle Ray taught me how to make my own—a method as simple as boiling an egg. The only problem is to get hold of the miniature cucumbers, usually available in the spring or early summer. They should average about 1½ inches [4 cm.] long and should be freshly picked and slightly underripe; try to get them with a tiny bit of the stem still attached. You should ideally have a stoneware crock in which to store them, but a Mason jar will also do. Make sure that the vinegar is of first quality and has an

acidity of at least 6 per cent. Otherwise, if the vinegar is too weak, the pickles may lose their crispness and may spoil.

If you are going to keep the pickles for a long time, it is a good idea to put a few fresh grape leaves into the crock with the pickles to keep them crisp.

To make about 3 quarts [3 liters]

3 lb.	baby pickling cucumbers (about 2½ quarts [2½ liters]), scrubbed, soaked overnight in heavily salted water, drained and dried	1½ kg.
about 1 quart	white wine vinegar	about 1 liter

Pack the cucumbers neatly and tightly into a 3-quart [3-liter] stoneware crock that will be their storage home. To determine how much vinegar is needed, pour it into the crock to about 1 inch [2½ cm.] above the top of the cucumbers. Now, pour off the vinegar into a large enameled, tinned or stainless-steel pan, and bring to a boil. Add an extra ½ cup [125 ml.] of vinegar. At once, pour the vinegar, still at a boil, over the cucumbers in the crock, cover the crock and let the cucumbers soak for a full 24 hours.

Again, pour off the vinegar into the pan, bring it back to a boil, adding an extra ½ cup of fresh vinegar, then pour the boiling vinegar over the cucumbers, cover, and let them soak for another 24 hours. Repeat exactly the same operation for the third time. This time, however, let the cucumbers soak, covered, in a cool place for six weeks. Take out the pickles as you need them. Always make sure that the remaining pickles are covered by liquid. Add more vinegar at any time, as necessary.

ROY ANDRIES DE GROOT
THE AUBERGE OF THE FLOWERING HEARTH

Sour Gherkins in Vinegar (Uncooked)

Cornichons au Vinaigre (à Cru)

To make about 2 quarts [2 liters]

2 lb.	small pickling cucumbers, rubbed to remove spikes (about 7 cups [1¾ liters])	1 kg.
1 cup	coarse salt	¼ liter
about 1½ quarts	wine vinegar	about 1½ liters
¼ lb.	small boiling onions	125 g.
6	dried hot red chilies	6
6	sprigs thyme	6
1	bay leaf, crumbled	1
4	sprigs tarragon	4
6	whole cloves	6

Mix the cucumbers with the salt, and leave in a bowl for 24 hours. Drain, wash the cucumbers in water mixed with

2 tablespoons [30 ml.] of vinegar, and drain again. Wipe them one by one. Arrange the cucumbers in a glass jar or in a stoneware pot, adding the onions, herbs and seasonings. Cover with wine vinegar. Close the jar or pot tightly, and refrigerate. Gherkins prepared in this way will be ready to eat in three to four weeks and will keep in excellent condition for months.

PROSPER MONTAGNÉ
THE NEW LAROUSSE GASTRONOMIQUE

Pickled Eggplant in the Mediterranean Style

This adds excitement to any antipasto; it can be served as a first course, with or without lettuce, and it's great in combination with hot or cold slices of beef, poultry, pork, veal or lamb. Add it to chef's salads; place it next to grilled hamburgers; include a large plateful in your next buffet.

To make about 3 quarts [3 liters]

4 lb.	eggplants, peeled and cut into julienne (about 3 quarts [3 liters])	2 kg.
4 tbsp.	salt	60 ml.
4 cups	white vinegar	1 liter
2 tbsp.	sugar	30 ml.
½ cup	olive oil	125 ml.
2	fresh hot chilies, stemmed, seeded and finely chopped	2
½ cup	finely cut chives	125 ml.
2 tbsp.	oregano	30 ml.
4	garlic cloves, halved	4

Put the eggplants in a colander, salt, toss and set them aside, over a bowl or in the sink, to drain for 30 minutes.

In a large saucepan, bring the vinegar and sugar to a boil. Add the eggplants and boil for five minutes. (Do not overcook.) Drain, reserving the cooking liquid.

Add the oil, chilies, chives and oregano to the eggplants. Toss well. Fill the jars, leaving ½ inch [1 cm.] of space at the top. Press the eggplant mixture to the bottom of the jar to release its juice. If more juice is needed to cover the eggplant, add 1 tablespoon [15 ml.] of the reserved cooking liquid. Distribute the garlic pieces among the jars and cover. Process *(pages 30-31)*. The pickled eggplant should be stored in a cool, dark place for four weeks before it is used.

JOE FAMULARO AND LOUISE IMPERIALE
THE FESTIVE FAMULARO KITCHEN

Eggplant in Oil

Melanzane Conciate

To make about 5 quarts [5 liters]

9 lb.	small, tender eggplants, stems left on, peeled (about 5 quarts [5 liters])	4 kg.
2 quarts	vinegar	2 liters
½ cup	salt	125 ml.
½	nutmeg, grated	½
10	black peppercorns	10
10	whole cloves	10
1	cinnamon stick, broken into pieces	1
6 cups	olive oil	1 ½ liters

Boil the eggplants in the vinegar with the salt. After 15 minutes, add the nutmeg, peppercorns and five of the cloves. Cook for another 15 minutes, or until tender. Drain thoroughly and cool. Place the eggplants in jars, distributing the remaining cloves and the cinnamon among them. Cover with the oil, close the jars and process *(pages 30-31)*.

EMMANUELE ROSSI (EDITOR)
LA VERA CUCINIERA GENOVESE

Mushrooms Preserved in Oil

Funghi sott'Olio

Small mushrooms are the most suitable for preserving in oil, as they are usually firm and blemish-free.

To make about 5 pints [2 ½ liters]

2 lb.	fresh mushrooms (about 3 quarts [3 liters])	1 kg.
2 cups	white wine vinegar	½ liter
	salt	
4	bay leaves, halved	4
6	whole cloves	6
1 tbsp.	white peppercorns	15 ml.
3 cups	olive oil	¾ liter

Place the mushrooms in an enameled, tinned or stainless-steel pan, cover them with the vinegar, and add a pinch of salt. Boil them until they begin to soften, about 10 minutes. Drain them thoroughly. Arrange the mushrooms in jars, placing the bay leaves, cloves and peppercorns between the layers. Cover the mushrooms with the olive oil and cover the jars. Process *(pages 30-31)*.

GIUSEPPE OBEROSLER (EDITOR)
IL TESORETTO DELLA CUCINA ITALIANA

Pickled Nasturtium Seeds

Choose seeds that have been gathered within seven days of the time the blossoms fell off. The pickled seeds can be chopped up and used instead of capers.

To make about 8 pints [4 liters]

2 quarts	nasturtium seeds	2 liters
	salt	
	fresh tarragon leaves	
	grated fresh horseradish	
Spiced vinegar		
2 quarts	white wine vinegar	2 liters
2	shallots, sliced	2
¼ cup	salt	50 ml.
3 tbsp.	white peppercorns	45 ml.
2 tbsp.	ground mace	30 ml.
2 tbsp.	grated nutmeg	30 ml.

Cover the nasturtium seeds with lightly salted water; let them remain for three days, changing the salt water daily. Drain them in a strainer and dry the seeds on a cloth.

Put the spiced vinegar ingredients into a saucepan. Boil over high heat for 10 minutes or so, then let the vinegar cool. Put the seeds into jars in layers, with a few tarragon leaves and plenty of grated horseradish between the layers. Strain the spiced vinegar over the seeds. Process *(pages 30-31)*.

FLORENCE WHITE
FLOWERS AS FOOD

Dilled Okra

This pickle is unusually crisp and amazingly good. Let it stand for three to four weeks before using, to allow the flavors to blend well.

To make 6 pints [3 liters]

3 lb.	young okra, each 4 inches [10 cm.] long	1½ kg.
6	heads fresh dill	6
1 quart	water	1 liter
2 cups	cider vinegar	½ liter
¼ cup	coarse salt	50 ml.
3 tsp.	celery seeds	15 ml.
6	garlic cloves	6

Pierce each okra with a fork; do not stem or slice it. Combine the water, vinegar and salt in a saucepan, and bring to a boil.

Divide the celery seeds, garlic and dill among six hot 1-pint [½-liter] jars. Pack the okra lengthwise into the jars.

Pour the boiling brine over the okra, filling the jars to within ½ inch [1 cm.] of the rims. Wipe the rims; adjust the lids.

Process the jars in a boiling water bath for 10 minutes. Start to count the processing time when the water in the bath returns to boiling. Remove the jars and complete the seals, unless the closures are self-sealing.

NELL B. NICHOLS AND KATHRYN LARSON (EDITORS)
FARM JOURNAL'S FREEZING & CANNING COOKBOOK

Pickled Okra

This tender-crisp pickle, peppery and tart, is a surprise to those who have only tasted okra as a cooked (and pleasantly bland and gelatinous) vegetable.

To make 6 pints [3 liters]

2 lb.	young okra pods, any darkened stem tips cut off, each pod left intact with a stub of stem (about 2 quarts [2 liters])	1 kg.
3	small dried hot chilies, halved and seeded, or ¾ tsp. [4 ml.] dried red pepper flakes	3
3	garlic cloves, halved	3
1 tbsp.	mustard seeds	15 ml.
1½ tsp.	dill seeds	7 ml.
2 cups	cider vinegar	½ liter
2 cups	rice vinegar or white wine vinegar	½ liter
2 cups	water	½ liter
5 tbsp.	coarse salt	75 ml.

Into each of six clean 1-pint [½-liter] jars, put half of a dried chili (or ⅛ teaspoon [½ ml.] of red pepper flakes), half of a garlic clove, ½ teaspoon [2 ml.] of the mustard seeds and ¼ teaspoon [1 ml.] of the dill seeds. Pack the okra into the jars, standing the pods upright and setting the stem ends and tips upward alternately; pack the pods just firmly enough to keep them upright.

In an enameled, tinned or stainless-steel pan, bring the vinegars, water and salt to a boil, and fill the jars with the boiling liquid to within ½ inch [1 cm.] of the rims. Wipe the rims and put on the lids. Put the jars on a rack in a deep kettle half-full of boiling water, and add more boiling water to cover the lids by 2 inches [5 cm.]. Bring to a boil, cover the pan and boil for 10 minutes.

Remove the jars from the water bath and let them cool. Let the pickles mellow for about a month before serving.

HELEN WITTY AND ELIZABETH SCHNEIDER COLCHIE
BETTER THAN STORE-BOUGHT

Oiled Onions

To make 8 pints [4 liters]

6 lb.	small boiling onions (about 4 quarts [4 liters])	2¾ kg.
1 cup	coarse salt	¼ liter
7 cups	water	1¾ liters
2 quarts	white vinegar	2 liters
2 cups	sugar	½ liter
¼ cup	pickling spices, tied in a muslin bag	50 ml.
¼ cup	olive oil	50 ml.

Scald the onions in boiling water for two minutes; drain. Cool in cold water and peel them. Cover with brine made by stirring the salt into the water, and let the onions stand overnight. The next day, drain the onions and wash them in cold water. Combine the vinegar, sugar and spice bag in a pan, and boil for three to five minutes. Remove the spice bag and add the onions. Let the mixture come just to a boil and pour at once into jars, filling them almost to the top. Pour in oil to fill the jars. Cover and process the onions *(pages 30-31)*. After the jars cool, shake them gently to distribute the oil.

RUBY CHARITY STARK GUTHRIE & JACK STARK GUTHRIE
A PRIMER FOR PICKLES

Pickled Onions

Hillosipulit

To make 3 to 4 pints [1 ½ to 2 liters]

1½ lb.	small boiling onions, peeled	¾ kg.
2½ cups	water	625 ml.
¼ cup	coarse salt	50 ml.
2½ cups	white vinegar	625 ml.
⅔ cup	brown sugar	150 ml.
¼ tsp.	ground mace	1 ml.
½-inch	slice fresh ginger or ¼ tsp. [1 ml.] ground ginger	1-cm.
10	white peppercorns	10

Bring the water and salt to a boil, and pour over the onions. Let stand for one day or overnight. Drain. Combine the remaining ingredients in a saucepan. Bring to a boil, cover, and simmer for three minutes. Drop in the onions, cover, and simmer for five minutes, or until the onions are half-done, transparent but still firm and crisp.

With a slotted spoon, remove the onions from the cooking liquid and place them in jars, filling the jars about two thirds full. Pour the cooking liquid through a sieve into the jars to

fill them. Cap the jars and process *(pages 30-31)*. Keep the onions for at least four or five days before serving.

ULLA KÄKÖNEN
NATURAL COOKING THE FINNISH WAY

Shepherd's Market Onion Rings

You will find these onion rings an interesting accompaniment to roasts and chops, and a wonderful addition to an antipasto platter.

To make 12 cups [3 liters]

3 lb.	Spanish onions	1½ kg.
3 cups	sugar	¾ liter
2 cups	vinegar	½ liter

Cut the onions into very thin slices and put them in a large nonmetal bowl, separating the slices into rings. Pour the sugar and vinegar over the onion rings and mix well. Cover, and let steep at room temperature for at least 12 hours (24 is better yet), stirring occasionally. The volume will reduce as the raw onions wilt and soften. Cover with plastic wrap and refrigerate. The onion rings will keep for two or three weeks.

CAROL CUTLER
HAUTE CUISINE FOR YOUR HEART'S DELIGHT

Pickled Pumpkin

To make about 5 pints [2 ½ liters]

4 to 5 lb.	pumpkin, halved, seeded, peeled and cubed (about 2½ quarts [2½ liters])	2 to 2½ kg.
	salt	
2 cups	vinegar	½ liter
1 cup	maple syrup	¼ liter
2	whole cloves	2
¼ tsp.	ground ginger	1 ml.
6	peppercorns	6
1	bay leaf	1
1	garlic clove, peeled and chopped	1

Sprinkle the pumpkin liberally with salt and let it stand for two to three hours. Boil together the remaining ingredients for about 10 minutes to make a clear syrup. Remove the bay leaf. Rinse the pumpkin, drain it thoroughly, pack it into hot jars, and pour the syrup over it. Close the jars and process *(pages 30-31)*. Store for at least two weeks before opening.

LEONARD LOUIS LEVINSON
THE COMPLETE BOOK OF PICKLES & RELISHES

Sweet-and-Sour Pumpkin

Kürbis Süss-Sauer

Winter squash —Hubbard, butternut or the like —can be substituted for the pumpkin.

To make about 4 ½ quarts [4 ½ liters]

6 to 7 lb.	pumpkin, halved, seeded, cut into chunks and peeled	2 ¾ to 3 kg.
4 cups	wine vinegar	1 liter
1 quart	water	1 liter
6 cups	sugar	1 ½ liters
2 tsp.	grated lemon peel	10 ml.
10	whole cloves	10
1	cinnamon stick	1
2 tbsp.	chopped fresh ginger	30 ml.
20	black peppercorns	20
1 tbsp.	salt	15 ml.

Boil the vinegar, water, sugar, lemon peel and spices together for 15 minutes. Add the pumpkin and simmer for 20 minutes, or until the pumpkin pieces are tender and transparent. Cover and let stand overnight. The next day, remove the pumpkin and pack it into jars. Boil the juice for five minutes, until it forms a thick syrup. Strain and pour over the pumpkin. Cover the jars. Process *(pages 30-31)*.

DOROTHEE V. HELLERMANN
DAS KOCHBUCH AUS HAMBURG

Pickled Peppers or Chilies

To make 8 pints [4 liters]

2 ½ lb.	red or green sweet peppers or fresh hot chilies (about 4 quarts [4 liters])	1 ¼ kg.
1 quart	vinegar	1 liter
1 quart	water	1 liter
4 tsp.	salt	20 ml.
	olive oil (optional)	

Wash the peppers or chilies thoroughly. Cut around the stem of each large pepper, and pull out the stem with the attached core and seeds. Leave the peppers whole or cut them into sections or strips, as desired. Hot chilies may be stemmed and seeded, or left whole with stems intact. Make two small slits in whole peppers or chilies.

Mix the vinegar and water; heat to a simmer—150° to 160° F. [65° to 70° C.]. The vinegar should not be allowed to boil. Add the salt.

Pack the peppers or chilies rather tightly into jars. Pour the hot vinegar mixture over the peppers to within ½ inch [1 cm.] of the jar rims. Or, if you wish to add olive oil, pour the vinegar mixture to within ¾ inch [2 cm.] of the rims and pour in the oil to within ½ inch of the rims. The peppers or chilies will be coated with oil when they pass through the oil layer as you use them.

Cover, and process for 15 minutes in a boiling water bath.

UNIVERSITY OF CALIFORNIA COOPERATIVE EXTENSION
SAFE METHODS FOR PREPARING PICKLES, RELISHES AND CHUTNEYS

Red Peppers Stuffed with Cabbage

Paprika mit Kohl Gefüllt

To make about 2 quarts [2 liters]

6 to 8	sweet red peppers	6 to 8
1	medium-sized cabbage, trimmed, quartered, cored and finely shredded	1
¼ cup	salt	50 ml.
1 quart	vinegar	1 liter
1 oz.	fresh horseradish, peeled and thinly sliced (about ¼ cup [50 ml.])	30 g.
3 tbsp.	black peppercorns	45 ml.
3 tbsp.	mustard seeds	45 ml.
2 tbsp.	sugar	30 ml.

Mix the cabbage with the salt and let stand overnight. The next day, squeeze out any liquid and mix the cabbage with 1 ¼ cups [300 ml.] of the vinegar.

Cut around the stem ends of the peppers, and pull out the stems and the attached seeds. Drain the cabbage and stuff it lightly into the peppers. Layer the peppers, stem ends upward, in 1-quart [1-liter] canning jars, placing a few horseradish slices, peppercorns and mustard seeds among the peppers. As you proceed, cover each pepper with a slice of horseradish, topped with a little of the remaining cabbage.

Add the sugar to the remaining vinegar, bring the mixture to a boil and use it to fill the jars. Close the jars and process the peppers *(pages 30-31)*.

ELEK MAGYAR
KOCHBUCH FÜR FEINSCHMECKER

Pickled Tomatoes

To make about 2 quarts [2 liters]

12	medium-sized green tomatoes (about 2 quarts [2 liters])	12
3 cups	water	¾ liter
1¼ cups	vinegar	300 ml.
	Pickle syrup	
⅔ cup	vinegar	150 ml.
2½ cups	water	625 ml.
5 cups	sugar	1¼ liters
10	whole cloves	10
¼	dried hot chili	¼

Prick the tomatoes all over with a fork. Place them in a large bowl. Mix the water and vinegar together, bring to a boil, pour the boiling liquid over the tomatoes and let the tomatoes stand for 24 hours.

To make the pickle syrup, boil all the syrup ingredients together until clear—about 15 minutes. Drain the tomatoes, add them to the syrup and simmer until they are tender. Remove the tomatoes and put them into jars. If the syrup has not thickened, boil it a little longer before pouring it over the tomatoes. Close the jars. Process *(pages 30-31)*.

INGA NORBERG
GOOD FOOD FROM SWEDEN

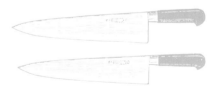

Pickled Turnips

Kabees el Lift

To make about 1 quart [1 liter]

1½ lb.	small white turnips (about 1 quart [1 liter])	¾ kg.
1	small beet, peeled	1
2 cups	water	½ liter
1 cup	vinegar	¼ liter
3	garlic cloves	3
2 tsp.	salt	10 ml.

Cut a slice from the tops and bottoms of the turnips. Cut each turnip lengthwise into ¼-inch [6-mm.] slices to within ½

inch [1 cm.] of the broad base of the turnip. Do not separate the slices entirely from each other. Soak the turnips in water to cover overnight. Wash them well in the morning. Place the turnips in a glass jar with the beet to give it color.

Combine the remaining ingredients with 2 cups [½ liter] of water, and boil them for 10 minutes to make a pickling solution. Let the solution cool completely, then pour it over the turnips. Refrigerate or process *(pages 30-31)*. The turnips will be ready for eating after three days.

LEONARD LOUIS LEVINSON
THE COMPLETE BOOK OF PICKLES AND RELISHES

To Pickle Walnuts

Walnuts are at the correct degree of ripeness for this recipe—fully developed, but still soft enough to be easily pierced with a needle—from about the end of June to mid-July in Northern states and from about the end of May to mid-June in the South and Southwest.

To make about 2 pints [1 liter]

2 lb.	green walnuts (about 1 quart [1 liter])	1 kg.
1 cup	salt	¼ liter
2	garlic cloves	2
4	shallots	4
2 tbsp.	mustard seeds	30 ml.
½ tsp.	grated nutmeg	2 ml.
1	blade mace (or substitute ¾ tsp. [4 ml.] ground mace)	1
4	whole cloves	4
8	black peppercorns	8
1 tsp.	ground ginger	5 ml.
2½ cups	distilled malt vinegar	625 ml.

Cut a hole in each walnut shell and pick out the nut. Place the nuts in a large pan of water, and let them simmer briskly for 45 minutes.

Dissolve the salt in 7½ cups [1¾ liters] of water to make a brine. Drain the walnuts, add them to the brine and let them soak for nine days.

In a mortar, pound together the garlic, shallots, mustard seeds, nutmeg, mace, cloves, peppercorns and ginger to make a paste. Add this paste to the vinegar and boil for about five minutes. Cool. Drain the walnuts and pack them in jars. Pour over the vinegar mixture to cover them and seal the jars. Store in refrigerator. The walnuts are ready for use after one month.

ANN BLENCOWE
THE RECEIPT BOOK OF ANN BLENCOWE

Chutneys and Relishes

Apricot and Cantaloupe Chutney

To make about 6 pints [3 liters]

1 lb.	dried apricots, cut into ½-inch [1-cm.] pieces	½ kg.
3	medium-sized cantaloupes, quartered, seeded, peeled and cut into ½-inch [1-cm.] cubes	3
1	fresh hot chili, stemmed, seeded and diced, or ½ tsp. [2 ml.] red pepper flakes	1
2 cups	dried currants	½ liter
1 tsp.	ground cloves	5 ml.
1 tsp.	grated nutmeg	5 ml.
2 tbsp.	salt	30 ml.
2 tbsp.	mustard seeds	30 ml.
¼ cup	finely chopped fresh or candied ginger	50 ml.
3	large garlic cloves, finely chopped	3
4½ cups	cider vinegar	1⅛ liters
1 lb.	dark brown sugar	½ kg.
4	medium-sized onions, finely chopped	4
½ cup	fresh orange juice	125 ml.
2 tbsp.	finely chopped fresh orange peel	30 ml.

Wash your preserving jars, and keep them in hot water until you are ready to fill them.

In a bowl, mix the apricots, chili, currants, cloves, nutmeg, salt, mustard seeds, ginger and garlic. In an enameled, tinned or stainless-steel pan, combine the vinegar and brown sugar, and bring to a boil over medium heat. Then add the mixture in the bowl to the saucepan. Bring this mixture to a gentle simmer and cook, uncovered, for 45 minutes.

Meanwhile, combine the onions, cantaloupes, and orange juice and peel in a bowl. Add the cantaloupe mixture to the apricot mixture and simmer for another 45 minutes.

Fill the jars and cap them. Store the jars in the refrigerator for about six weeks before using the chutney. Or process to keep for a year *(pages 30-31)*.

JOE FAMULARO AND LOUISE IMPERIALE
THE FESTIVE FAMULARO KITCHEN

Apricot Chutney

To make apple chutney in the same way, use 2 quarts [2 liters] of peeled, cored apples, substituting almonds for the walnuts.

To make about 4 pints [2 liters]

1 lb.	dried apricots or 6 cups [1½ liters] fresh apricots, pitted	½ kg.
4	medium-sized onions, sliced	4
1⅛ cups	seedless raisins	275 ml.
2½ cups	white wine vinegar	625 ml.
1 lb.	dark brown sugar	½ kg.
4 tbsp.	salt	60 ml.
1 cup	preserved ginger	¼ liter
1 tbsp.	mustard seeds	15 ml.
1 tsp.	cayenne pepper	5 ml.
½ tsp.	ground turmeric	2 ml.
1	orange, the peel grated and the juice strained	1
½ cup	walnuts	125 ml.

Put all of the ingredients except the walnuts into a pan, and cook gently to a soft mash, about one and one half hours. Add the walnuts. Pack into jars and cover. Process *(pages 30-31)*. Keep the chutney for at least one month before using it.

JANE GRIGSON
THE ART OF MAKING SAUSAGES, PÂTÉ AND OTHER CHARCUTERIE

A Banana Chutney

To make about 7 pints [3½ liters]

16	ripe bananas, peeled and sliced	16
6	large onions, finely chopped	6
1 lb.	dates, pitted and finely chopped	½ kg.
2 cups	finely chopped candied ginger	½ liter
1 tbsp.	salt	15 ml.
3 tbsp.	pickling spices, tied in a cloth bag	45 ml.
1 quart	vinegar	1 liter
1½ cups	molasses	375 ml.

Put all the ingredients except the vinegar and molasses into an enameled, tinned or stainless-steel pan. Cover with the vinegar. Boil for five minutes. Take out the spice bag. Add the molasses and continue to cook in a 350° F. [180° C.] oven, or over low heat, until the chutney is thick and brown, about three hours. Put into jars, cover and process *(pages 30-31)*.

PEGGY HUTCHINSON
GRANDMA'S PRESERVING SECRETS

Apple-Ginger Chutney

To make about 7 pints [3 ½ liters]

5 lb.	large firm apples, cored and coarsely chopped (about 5 quarts [5 liters])	2½ kg.
¾ cup	preserved ginger in syrup, finely chopped, syrup reserved	175 ml.
3 cups	brown sugar	¾ liter
4 cups	cider vinegar	1 liter
3 tbsp.	salt	45 ml.
1 tsp.	ground cloves	5 ml.
1 tsp.	ground allspice	5 ml.
3 tsp.	ground ginger	15 ml.
2	large green peppers, halved, seeded, deribbed and finely chopped	2
1	large onion, finely chopped	1
2 cups	dried currants	½ liter
1	lemon, the peel grated, the juice strained	1

In a large enameled, tinned or stainless-steel pan, bring to a boil the sugar, vinegar, salt and spices. Add the peppers, onion and currants, and simmer for 30 minutes. Add all the other ingredients and boil gently for 30 minutes, stirring frequently. Put into jars, cover and process *(pages 30-31)*.

BLANCHE POWNALL GARRETT
CANADIAN COUNTRY PRESERVES & WINES

Sweet Mango Chutney

To make about 4 pints [2 liters]

3 lb.	green mangos, halved, pitted, peeled and diced (about 9 cups [2¼ liters])	1½ kg.
	salt	
2	garlic cloves, coarsely chopped	2
⅓ cup	coarsely chopped fresh ginger	75 ml.
1 tbsp.	cayenne pepper	15 ml.
2 cups	vinegar	½ liter
6 cups	sugar	1½ liters
1 lb.	plums, scalded, peeled, halved, pitted and chopped (about 2½ cups [625 ml.])	½ kg.

Sprinkle the mangos with salt and set aside for one hour. Drain the mangos.

Grind the garlic, ginger and cayenne pepper in a mortar with ⅔ cup [150 ml.] of the vinegar to make a smooth paste. Put the mangos and sugar in a saucepan, and boil together until the mangos are tender and the mixture forms a thick jam, about 45 minutes. Add the seasoning paste and stir gently. Add the plums and cook, stirring, for eight minutes, or until the plums are soft. Add the remaining vinegar, and more salt if necessary, and simmer for 15 minutes, or until thick and dry. Put in jars, cover and process *(pages 30-31)*.

PREMILA LAL
PREMILA LAL'S INDIAN RECIPES

Caribbean Mango Chutney

This is a close relative of Indian mango chutney, brought to the Caribbean by East Indian immigrants.

To make 3 pints [1 ½ liters]

12	mangos, halved, pitted, peeled and sliced	12
¾ cup	raisins	175 ml.
½ cup	pitted and chopped dates	125 ml.
1	sweet red pepper, halved, seeded, deribbed and diced	1
2 cups	cider vinegar	½ liter
¼ lb.	fresh ginger, peeled and sliced (about ½ cup [125 ml.])	125 g.
¼ cup	salt	50 ml.
1	garlic clove	1
1	onion, sliced (optional)	1
1 tbsp.	ground mustard seeds (optional)	15 ml.
1 tbsp.	celery seed (optional)	15 ml.
¼ cup	chopped fresh peanuts (optional)	50 ml.
½ cup	strained fresh lemon juice (optional)	125 ml.
1	large yellow summer squash, sliced (optional)	1

Combine the raisins, dates, sweet pepper and half of the vinegar, and let stand for 24 hours. Then add the remaining ingredients. Boil gently for 45 minutes, or until the mixture is thick. Ladle into jars, cover and process *(pages 30-31)*.

CONNIE AND ARNOLD KROCHMAL
CARIBBEAN COOKING

Mango Chutney

To make about 3 pints [1 ½ liters]

2 lb.	green mangos, halved, pitted, peeled and sliced (about 6 cups [1 ½ liters])	1 kg.
½ lb.	fresh ginger, peeled, half coarsely chopped and half thinly sliced	¼ kg.
3½ cups	red currants, stemmed	875 ml.
8 cups	sugar	2 liters
2 cups	vinegar	½ liter
3 tbsp.	cayenne pepper	45 ml.
1 tbsp.	salt	15 ml.

In a large mortar, grind the chopped ginger and half of the currants. Put all of the ingredients except the mangos into a saucepan. Stirring, cook for 15 minutes over medium heat. Add the mangos, and simmer until the mixture has a jam-like consistency and the mangos are tender, about 30 minutes. Put in jars, cover and process *(pages 30-31)*.

PREMILA LAL
PREMILA LAL'S INDIAN RECIPES

California Chutney

To make about 2 pints [1 liter]

1 cup	prunes	¼ liter
1½ cups	pear or cider vinegar	375 ml.
2 cups	brown sugar	½ liter
1 tsp.	dry mustard	5 ml.
½ tsp.	ground cinnamon	2 ml.
½ tsp.	salt	2 ml.
⅛ tsp.	cayenne pepper	½ ml.
3	medium-sized apples, peeled, cored and chopped	3
1 cup	seedless raisins	¼ liter
1 cup	chopped onions	¼ liter
2	medium-sized tomatoes, peeled, seeded and chopped	2

Cover the dried prunes with water and boil for 10 minutes. Drain; cut the fruit from the pits into small pieces. Combine the vinegar, sugar, mustard, cinnamon, salt and cayenne in an enameled or stainless-steel pan. Heat to boiling; add the prunes, apples, raisins, onions and tomatoes. Cover and boil, stirring frequently with a wooden or enameled spoon. When the mixture is of the desired consistency, in about 30 minutes, pour it into jars and cover. Process *(pages 30-31)*.

FLORENCE BROBECK
OLD-TIME PICKLING AND SPICING RECIPES

Peach-Date Chutney

To make about 3 ½ pints [1 ¾ liters]

2 lb.	ripe peaches, peeled, halved, pitted and sliced (about 2 quarts [2 liters])	1 kg.
2	medium-sized onions, chopped	2
3 cups	sugar	¾ liter
½ lb.	dates, chopped (about 1 ½ cups [375 ml.])	¼ kg.
½ cup	sliced candied ginger	125 ml.
1 cup	tarragon vinegar	¼ liter
½ cup	white vinegar	125 ml.

Mix all the ingredients except the peaches in a large enameled, tinned or stainless-steel pan. Cook for 30 minutes, stirring frequently. Add the peaches, and continue cooking for about 20 minutes more, or until well blended and thickened. Put in jars and cover. Process *(pages 30-31)*.

MRS. DON RICHARDSON (EDITOR)
CAROLINA LOW COUNTRY COOK BOOK OF GEORGETOWN,
SOUTH CAROLINA

Mild Green-Tomato Chutney

All of the vinegar may be added at the beginning of cooking; in this case, the cooking time will be about two hours.

If you have the time, peel the tomatoes. To do this, pour boiling water over them, leave them for a few minutes, and peel them with a sharp knife. This small extra trouble makes a perceptible difference to the finished chutney.

To make about 5 pints [2 ½ liters]

2 lb.	green tomatoes, chopped (about 4 cups [1 liter])	1 kg.
2 lb.	tart apples, peeled, cored and sliced (about 2 quarts [2 liters])	1 kg.
2	medium-sized onions, sliced	2
4 cups	seedless raisins	1 liter
4 cups	brown sugar	1 liter
2 tsp.	ground ginger	10 ml.
2 tsp.	ground allspice	10 ml.
2 tsp.	crushed black peppercorns	10 ml.
2 tbsp.	salt	30 ml.
2	garlic cloves, coarsely chopped	2
3½ cups	Orléans or other white wine vinegar	875 ml.

Put all the ingredients except the vinegar into an enameled, tinned or stainless-steel pan. Moisten with a little of the vinegar; cook gently, adding the rest of the vinegar as the chutney boils down, for about an hour. Toward the end, stir constantly; as they thicken, chutneys stick very easily. When the mixture is of a jamlike consistency it is ready. Pour into jars, cover and process *(pages 30-31)*.

ELIZABETH DAVID
SPICES, SALT AND AROMATICS IN THE ENGLISH KITCHEN

Green-Tomato Chutney

To make spiced vinegar, heat 2½ cups [625 ml.] of vinegar with a cinnamon stick, five blades of mace, and 1 teaspoon [5 ml.] each of whole cloves, whole allspice and black peppercorns. When the mixture is almost at the boiling point, remove it from the heat and let it cool. Strain the vinegar and use it immediately or store it in covered bottles.

To make about 5 pints [2 ½ liters]

4 lb.	small green tomatoes, very thinly sliced (about 2 quarts [2 liters])	2 kg.
	salt	
1 lb.	tart apples, peeled, cored and finely chopped (about 4 cups [1 liter])	½ kg.
1¼ lb.	shallots, finely chopped (about 3 cups [¾ liter])	550 g.
1 tsp.	celery salt	5 ml.
6	dried hot chilies, tied in a cheesecloth bag	6
1 lb.	brown sugar	½ kg.
2½ cups	cider vinegar, preferably spiced	625 ml.
4	medium-sized ripe red tomatoes, peeled and chopped	4
2	sweet red or green peppers, broiled, skinned, halved, seeded, deribbed and cut into short strips	2

Sprinkle the green tomatoes with salt, and let them drain in a colander for an hour or two. Rinse them. Put all of the ingredients except the broiled peppers and ripe tomatoes into a pan, and bring to a boil. Simmer for 15 minutes, or until most of the excess liquid has evaporated, then add the broiled peppers and ripe tomatoes. Simmer until thick, about one hour. Remove the chilies and pour the mixture into clean jars. Process *(pages 30-31)*.

CAROLINE CONRAN
BRITISH COOKING

Fruit Mustard Pickles

Mostarda Casalinga

Suitable fruits for this pickle include apricots, cherries, citrons, small figs, honeydew melon or cantaloupe, peaches or plums. The fruit should be peeled and pitted, if necessary, and cut into pieces.

To make about 2 quarts [2 liters]

2 quarts	prepared fruit	2 liters
½ cup	dry mustard	125 ml.
3½ cups	sugar	875 ml.
1 cup	water	¼ liter
⅔ cup	white wine vinegar	150 ml.

Combine 2½ cups [625 ml.] of the sugar with the water, and cook over medium heat for about 15 minutes, or until the sugar has completely dissolved.

Add the fruit and cook over low heat for 10 to 20 minutes, until the fruit begins to soften. Remove from the heat and pour into a bowl.

Combine the remaining sugar with the vinegar, and simmer for 15 minutes to make a thick syrup. Let it cool. Stir in the mustard and let the mixture rest for one hour.

Add the mustard syrup to the cooked fruit, mix and pack in jars. Process *(pages 30-31)*.

GIUSEPPE OBEROSLER (EDITOR)
IL TESORETTO DELLA CUCINA ITALIANA

Rhubarb and Raisin Relish

To make spiced vinegar, heat 2½ cups [625 ml.] of cider vinegar with a cinnamon stick, five blades of mace, and 1 teaspoon [5 ml.] each of whole cloves, whole allspice and black peppercorns. When the mixture is almost at the boiling point, remove it from the heat and let it cool. Strain and use immediately or store in covered jars.

To make about 2 pints [1 liter]

1 lb.	rhubarb, leaves removed, stalks trimmed and cut into small pieces (about 4 cups [1 liter])	½ kg.
2 cups	raisins	½ liter
2	onions, chopped	2
1⅓ cups	brown sugar	325 ml.
1¼ cups	cider vinegar, preferably spiced	300 ml.
⅔ cup	water	150 ml.
½ tsp.	whole allspice	2 ml.
½ tsp.	salt	2 ml.
½ tsp.	whole cloves	2 ml.
1 tsp.	mustard seeds	5 ml.
¼ tsp.	celery seeds	1 ml.

Combine the sugar, vinegar, water, spices and salt in an enameled, tinned or stainless-steel pan. Bring to a boil, and boil for five minutes. Add the rhubarb and the onions, and cover the pan. Simmer gently for 45 minutes. Add the raisins and, stirring from time to time, cook uncovered for about one hour, until the mixture is thick. Pour the relish into clean hot jars and process *(pages 30-31)*.

CAROLINE CONRAN
BRITISH COOKING

Grannie's Rhubarb Relish

To make about 6 pints [3 liters]

2 lb.	rhubarb, leaves removed, stalks trimmed and chopped (about 4 cups [1 liter])	1 kg.
4 cups	chopped onions	1 liter
2 cups	cider vinegar	½ liter
5 cups	brown sugar	1¼ liters
1 tsp.	ground cinnamon	5 ml.
½ tsp.	ground cloves	2 ml.
1 tsp.	ground allspice	5 ml.
½ tsp.	ground black pepper	2 ml.
2 tsp.	salt	10 ml.

Mix all the ingredients together in a large enameled, tinned or stainless-steel pan, and boil, uncovered, over medium heat, stirring frequently, for about 45 minutes, or until the mixture forms a thick purée, with all the excess liquid absorbed. Pour the relish into hot jars and process *(pages 30-31)*. It is ready for use after a month.

BRITISH COLUMBIA WOMEN'S INSTITUTES
ADVENTURES IN COOKING

Corn Salad

To make 3 pints [1 ½ liters]

10	ears of corn, kernels cut off cobs (about 5 cups [1¼ liters] kernels)	10
1	green pepper, halved, seeded, deribbed and cut into small pieces	1
½	red pepper, halved, seeded, deribbed and cut into small pieces	½
2	onions, chopped	2
½ tsp.	celery seed	2 ml.
1½ tsp.	dry mustard	7 ml.
1⅓ cups	vinegar	325 ml.
1⅓ cups	sugar	325 ml.
⅓ tsp.	ground turmeric	2 ml.

Mix the corn with the other ingredients, and boil all together slowly for one half hour, or until the vegetables are tender. Ladle into jars. Process *(pages 30-31)*.

RUTH HUTCHISON
THE PENNSYLVANIA DUTCH COOKBOOK

Green-Tomato Relish

To make about 6 pints [3 liters]

12 lb.	green tomatoes, blanched, peeled and thinly sliced (about 5½ quarts [5½ liters])	5½ kg.
4	onions, chopped	4
4	green peppers, halved, seeded, deribbed and finely chopped	4
½ cup	salt	125 ml.
2 tbsp.	whole cloves	30 ml.
2 tbsp.	pickling spices	30 ml.
2 tbsp.	black peppercorns	30 ml.
½ cup	mustard seeds, coarsely crushed in a mortar	125 ml.
1 lb.	brown sugar	½ kg.
7½ cups	wine vinegar	1¾ liters

Combine all of the ingredients in an enameled, tinned or stainless-steel pan. Stirring occasionally, heat the mixture to simmering and cook over low heat for three hours. Pour into jars, cover and process *(pages 30-31)*.

ELIZABETH SCHULER
MEIN KOCHBUCH

Zucchini and Green-Chili Relish

To make 4 pints [2 liters]

2½ lb.	ground or finely chopped zucchini (about 2½ quarts [2½ liters])	1¼ kg.
3 to 5	fresh green chilies, stemmed, and ground or finely chopped	3 to 5
5	large onions, ground or finely chopped	5
5 tbsp.	coarse salt	75 ml.
2¼ cups	cider vinegar	550 ml.
4 cups	sugar	1 liter
1 tbsp.	dry mustard	15 ml.
1 tbsp.	ground turmeric	15 ml.
1 tbsp.	grated nutmeg	15 ml.
2 tsp.	celery seeds	10 ml.
1 tbsp.	cornstarch	15 ml.
1	fresh red chili, stemmed, and ground or finely chopped, or 1 tsp. [5 ml.] coarsely ground dried red chili	1

In an enameled, tinned or stainless-steel pan, mix the zucchini, green chilies and onions with the salt, and let the mixture stand in the refrigerator overnight. Drain, and rinse with cold water; drain, rinse and drain the vegetables again. Mix with the remaining ingredients and boil for 30 minutes. Pour into hot jars, leaving ½ inch [1 cm.] headspace. Process *(pages 30-31)*.

PATRICIA GINS (EDITOR)
"NEW MEXICO PRIZED RECIPES" GREAT SOUTHWEST COOKING CLASSIC

Preserved Garden Vegetables

Giardiniera Delicata

There should be an equal volume —about 2 cups [½ liter]—of each of the prepared vegetables.

To make about 7 pints [3½ liters]

4	small carrots, cut into small pieces	4
½ lb.	green beans, cut into small pieces	¼ kg.
4	celery ribs, cut into small pieces	4
½	small cauliflower, cored and cut into small pieces	½
½ lb.	fennel, cut into small pieces	¼ kg.
2 cups	small boiling onions	½ liter
2	small sweet green peppers, halved, seeded, deribbed and cut into small pieces	2
1 quart	white wine vinegar	1 liter
⅓ cup	olive oil	75 ml.
½ cup	salt	125 ml.
½ cup	sugar	125 ml.

Combine the vinegar, oil, salt and sugar in a large enameled, tinned or stainless-steel pan, and bring to a boil. Add the vegetables, beginning with the hardest (the carrots), then add the green beans and celery, and after a few minutes, the cauliflower, fennel and onions. Allow no more than 20 minutes from the time the carrots are put in until the end of the cooking time. In this way the vegetables will remain crunchy. Just before removing the pan from the heat, add the peppers, but do not allow them to boil. Put the mixture into jars. Process *(pages 30-31)*.

GIANNA MONTECUCCO ROGLEDI
SOTTO VETRO

Hot Dog Relish

To make about 5 pints [2 ½ liters]

3	large onions, coarsely chopped	3
¼	head cabbage, coarsely chopped	¼
6	sweet green peppers, halved, seeded, deribbed and cut into small pieces	6
3	sweet red peppers, halved, seeded, deribbed and cut into small pieces	3
5	medium-sized tomatoes, quartered	5
¼ cup	coarse salt	50 ml.
1 quart	white or cider vinegar	1 liter
1 cup	water	¼ liter
3 cups	sugar	¾ liter
¾ tsp.	ground turmeric	4 ml.
1 tbsp.	dry mustard	15 ml.
1 tbsp.	mustard seeds	15 ml.
½ tbsp.	celery seed	7 ml.
about 1 quart	dill pickles, chopped	about 1 liter
about 1 quart	sweet pickles, chopped	about 1 liter

Put the raw vegetables through a food grinder with a coarse disk. Sprinkle them with the salt and let stand overnight; rinse the vegetables and drain them. Combine the vinegar, water, sugar and spices, and pour the mixture over the vegetables. Bring to a boil in an enameled, tinned or stainless-steel pan and cook for three minutes. Add the desired amounts of dill pickles and sweet pickles to the mixture just before pouring it hot into jars. If you wish, add some of the liquid from the pickle jars. Cover and process *(pages 30-31)*.

RUBY CHARITY STARK GUTHRIE & JACK STARK GUTHRIE
A PRIMER FOR PICKLES

Lindberg Relish

If you do not have a cool cellar in which to store this relish, pack it into jars and process it as explained on pages 30-31. White mustard seeds are milder than the more common yellow seeds, but yellow mustard seeds can be used instead.

	To make about 9 quarts [9 liters]	
2	large white cabbages, halved, cored, and finely chopped	2
8	carrots, finely chopped	8
12	onions, finely chopped	12
2	sweet red peppers, halved, seeded, deribbed and finely chopped	2
2	sweet green peppers, halved, seeded, deribbed and finely chopped	2
⅔ cup	salt	150 ml.
2 quarts	malt vinegar	2 liters
2 tbsp.	white mustard seeds	30 ml.
1 tbsp.	celery seed	15 ml.
6 cups	sugar	1 ½ liters

Mix the vegetables with the salt, then place in a large colander to drain overnight. Discard the drained liquid and put the vegetables in a crock. Stir in the remaining ingredients and tightly cover the crock.

This relish does not require cooking or sealing if it is kept in a cool (40° to 55° F. [5° to 15° C.]) place. It can be used in about a week and will keep for several months, if the vinegar always covers the contents of the crock. Or put the relish into jars and process *(pages 30-31)*.

BRITISH COLUMBIA WOMEN'S INSTITUTES
ADVENTURES IN COOKING

Late-Summer or Early-Autumn Pickle

Achards de Fin d'Été, Début Automne

To make about 7 pints [3 ½ liters]

about 1 lb. each	small pickling cucumbers, small green tomatoes, small carrots, small boiling onions, shallots, cauliflower florets and small green beans	about ½ kg. each
	coarse salt	
	vinegar	
	fresh tarragon sprigs	
	small fresh red and green hot chilies	

Mix the cucumbers in a bowl with a handful of coarse salt. Turning them two to three times, leave the cucumbers at room temperature for 24 hours to allow them to exude their excess liquid.

Using a trussing needle, prick each tomato deeply in three or four places. Peel the carrots, onions and shallots, and slice the carrots into rounds about 1 inch [2½ cm.] thick. Bring a large pot of water to a boil, adding 1½ tbsp. [22 ml.] of coarse salt for each 4 cups [1 liter] of water. Throw in all of the vegetables except the cucumbers. Boil for three minutes, then drain the vegetables and rinse them in cold water. Drain them well.

Drain the cucumbers and wipe them. Measure all of the vegetables.

For each 1½ cups [375 ml.] of vegetables, use 1 cup [¼ liter] of vinegar and 1 teaspoon [5 ml.] of coarse salt. Bring the vinegar and salt to a boil in an enameled, tinned or stainless-steel pan, boil for five minutes, then cool.

Layer the vegetables in jars, adding three or four sprigs of tarragon and a few small hot chilies to each jar. Cover with the cooled vinegar, close the jars and leave at room temperature for a week. Strain off the vinegar into a saucepan, boil it for five minutes and let it cool. Pour the vinegar back into the jars, adding fresh vinegar if the vegetables are not covered. Close the jars and process *(pages 30-31)*. After opening a jar, refrigerate it and use the contents within a month.

CÉLINE VENCE
LE GRAND LIVRE DES CONSERVES, DES CONFITURES
ET DE LA CONGÉLATION

Chowchow

To make about 24 pints [12 liters]

2	green cabbages, quartered, cored and cut into small pieces	2
2	cauliflowers, cores removed, cut into small florets	2
1 lb.	small boiling onions	½ kg.
12	small pickling cucumbers	12
12	small tomatoes (about 3 lb. [1½ kg.])	12
6	bunches celery, leaves removed, chopped	6
8 quarts	vinegar	8 liters
1 cup	dry mustard	¼ liter
1 cup	mustard seeds	¼ liter
½ cup	Dijon mustard	125 ml.
½ cup	ground turmeric	125 ml.
¼ cup	ground cloves	50 ml.

Using a separate pan for each type of vegetable, boil the vegetables in water until they are tender. Drain. In an enameled, tinned or stainless-steel pan, mix the vinegar with the dry mustard, mustard seeds and Dijon mustard, the turmeric and cloves, and set this mixture on the heat. When

it boils, mix all the vegetables together in a large bowl and pour the vinegar mixture over them. Pack into jars. Process *(pages 30-31)*. Leave the relish for a month before using.

OSCAR TSCHIRKY
THE COOK BOOK BY "OSCAR" OF THE WALDORF

Mixed Vegetables in Tomato Sauce
Giardiniera al Pomodoro

To make about 5 pints [2½ liters]

2 lb.	tomatoes, cut into chunks (about 4 cups [1 liter])	1 kg.
2	medium-sized carrots, cut into small pieces	2
2	celery ribs, cut into small pieces	2
¼ lb.	green beans, cut into small pieces (about 1 cup [¼ liter])	125 g.
1	large green pepper, halved, seeded, deribbed and cut into small pieces	1
2	small zucchini, cut into small pieces	2
1 cup	small boiling onions	¼ liter
½ cup	olive oil	125 ml.
½ tbsp.	salt	7 ml.
1 tbsp.	sugar	15 ml.
2 cups	vinegar	½ liter
6	fresh sage leaves	6
1 cup	fresh basil leaves	¼ liter
¼ tsp.	grated nutmeg	1 ml.
3	small pickling cucumbers, cut into thick rounds	3
2 tbsp.	capers	30 ml.

Cook the tomatoes over low heat for about one hour, until they are reduced to a thick purée. Press them through a sieve or food mill, and return the tomatoes to the pan. Stir in the oil, salt and sugar, and 1¼ cups [300 ml.] of the vinegar, and bring the mixture to a boil.

Add the carrots, celery, green beans and onions, and cook for five minutes, or until the vegetables begin to soften. Add the peppers and zucchini, sage, basil and nutmeg, and cook for only a few minutes more. The vegetables must remain slightly crunchy.

Remove the pan from the heat. Bring the remaining vinegar to a boil and cook the cucumbers for 10 minutes, or until they begin to soften. Drain the cucumbers and add them to the vegetable mixture, with the capers. Put the mixture into jars, cover and process *(pages 30-31)*.

GIANNA MONTECUCCO ROGLEDI
SOTTO VETRO

Piccalilli

For this pickle, use a mixture of cauliflower, green beans, summer squash, small carrots and shallots. The shallots should be peeled but left whole, the other vegetables cut into small pieces.

To make about 4 quarts [4 liters]

4 lb.	prepared mixed vegetables (about 4 quarts [4 liters])	2 kg.
½ cup	salt	125 ml.
½ cup	dry mustard	125 ml.
¼ cup	ground turmeric	50 ml.
2 tbsp.	flour	30 ml.
7 cups	vinegar	1¾ liters

Sprinkle the vegetables with the salt and let stand for 24 hours. Then drain.

Mix the mustard, turmeric and flour to a smooth paste with a little of the vinegar. Pour this paste into the rest of the vinegar, then add the vegetables and bring slowly to a boil, stirring all the time. Boil, uncovered, for 30 minutes, put into jars and process *(pages 30-31)*. The piccalilli is ready for use in two or three weeks. The flavor improves with keeping.

THE DAILY TELEGRAPH
400 PRIZE RECIPES FOR PRACTICAL COOKERY

Vegetables Preserved in Vinegar for the Winter

Légumes pour l'Hiver Confits dans le Vinaigre

Refrigerated, the vegetables will keep for several weeks.

To make about 2 quarts [2 liters]

1¼ lb.	green beans, cut into 2-inch [5-cm.] julienne (about 3 cups [¾ liter])	⅔ kg.
3	medium-sized carrots, cut into 2-inch [5-cm.] julienne	3
5	celery ribs, cut into 2-inch [5-cm.] julienne	5
1	medium-sized green pepper, halved, seeded, deribbed and cut into 2-inch [5-cm.] julienne	1
10 oz.	small boiling onions (about 1½ cups [375 ml.])	300 g.
1 quart	white wine vinegar	1 liter
½ cup	sugar	125 ml.
1 tbsp.	salt	15 ml.
1¼ cups	olive oil	300 ml.

In a large enameled, tinned or stainless-steel pan, bring the vinegar, sugar and salt slowly to a boil. Meanwhile prepare the vegetables. Then add the vegetables to the vinegar mixture and simmer for 10 minutes. Take the pan from the heat and let the vegetables cool. Fill jars with the vegetables and vinegar. When the jars are almost full, pour a thick layer of olive oil over the surface of the vinegar. Cover the jars, and process *(pages 30-31)* or refrigerate.

TINA CECCHINI
LES CONSERVES DE FRUITS ET LÉGUMES

Mixed Vegetables in Vinegar

Giardiniera Sotto Aceto

To make about 6 pints [3 liters]

10	small, tender carrots, cut into pieces	10
1	small cauliflower, cut into florets	1
2 cups	green beans, cut into pieces	½ liter
2 cups	small boiling onions, peeled	½ liter
1	small bunch celery, cut into pieces	1
1	sweet green pepper, halved, seeded, deribbed and cut into pieces	1
1	sweet red pepper, halved, seeded, deribbed and cut into pieces	1
1¼ cups	small pickling cucumbers	300 ml.
13 cups	white wine vinegar	3¼ liters
2 tsp.	salt	10 ml.
2½ cups	water	625 ml.
2 tbsp.	black peppercorns	30 ml.
2 tbsp.	juniper berries	30 ml.
1½ cups	olive oil	375 ml.

Put 2½ cups [625 ml.] of the vinegar into an enameled, tinned or stainless-steel pan with the water and the salt. Bring to a boil and in it cook the vegetables. Cook the carrots for 10 minutes, then remove them with a slotted spoon. Add the cauliflower and cook for seven to eight minutes. Remove. Cook the beans, onions and peppers in separate batches for just a few minutes each: They should remain crunchy. Discard the cooking liquid. Boil the cucumbers in 1¼ cups [300 ml.] of vinegar; drain them and reserve the vinegar.

Layer the vegetables in jars, scattering the peppercorns and juniper berries among them.

Add the remaining vinegar and the oil to the vinegar in which the cucumbers cooked. Bring to a boil. Pour it over the vegetables and close the jars. Process *(pages 30-31)*.

GIANNA MONTECUCCO ROGLEDI
SOTTO VETRO

Mild or Hot Piccalilli

Depending on what is available, any combination of vegetables may be used with either a mild or a hot sauce. Choose from florets of cauliflower, slices of zucchini or any other summer squash, small whole boiling onions or shallots, cubed cucumber, cut green beans, sliced celery and strips of green or red peppers.

To make about 6 pints [3 liters]

6 lb.	prepared mixed vegetables (about 3 quarts [3 liters])	3 kg.
2 cups	coarse salt	½ liter

Mild sweet sauce

1 tbsp.	ground turmeric	15 ml.
4 tsp.	dry mustard	20 ml.
4 tsp.	ground ginger	20 ml.
1 cup	sugar	¼ liter
7 cups	distilled vinegar	1¾ liters
½ cup	cornstarch	125 ml.

Hot sharp sauce

1 tbsp.	ground turmeric	15 ml.
3 tbsp.	dry mustard	45 ml.
8 tsp.	ground ginger	40 ml.
¾ cup	sugar	175 ml.
5 cups	distilled vinegar	1¼ liters
¼ cup	cornstarch	50 ml.

Layer the prepared vegetables with the salt in a large bowl, and cover with a plate to ensure that the vegetables really steep in the salt. Leave for 24 hours. Rinse the vegetables and drain them thoroughly.

Using the proportions for either the mild or hot sauce, put the turmeric, mustard, ginger, sugar and all but 3 tablespoons [45 ml.] of the vinegar in an enameled, tinned or stainless-steel pan. Mix well. Add the vegetables, bring to a boil, and simmer gently for 15 to 20 minutes, testing the firmness of the vegetables occasionally. The degree of crispness or tenderness of the vegetables is a matter for individual taste, but do not overcook them.

When ready, remove the vegetables with a perforated spoon and pack them into hot jars. Dissolve the cornstarch in the reserved vinegar and stir it into the sauce in the pan. Bring to a boil and boil for three minutes, stirring continuously. Pour the sauce over the vegetables and process *(pages 30-31)*. The piccalilli will be ready for use in six weeks.

OLIVE ODELL
PRESERVES AND PRESERVING

Mustard Pickles

To make about 3½ quarts [3½ liters]

1 lb.	small pickling cucumbers (about 2 cups [½ liter])	½ kg.
1 lb.	green beans, cut into ½-inch [1-cm.] lengths (about 4 cups [1 liter])	½ kg.
4	green peppers, halved, seeded, deribbed and cut into ½-inch [1-cm.] squares	4
4	red peppers, halved, seeded, deribbed and cut into ½-inch [1-cm.] squares	4
½ lb.	small boiling onions (about 2 cups [½ liter])	¼ kg.
1 lb.	green tomatoes, roughly chopped (about 2 cups [½ liter])	½ kg.
1	small cauliflower, cored and separated into florets	1
¼ lb.	shelled lima beans (about 1 cup [¼ liter])	125 g.
1⅓ cups	salt	325 ml.
¾ cup	flour	175 ml.
2 cups	sugar	½ liter
¼ cup	dry mustard	50 ml.
2 tsp.	ground turmeric	10 ml.
1 tbsp.	celery seed	15 ml.
½ tsp.	ground white pepper	2 ml.
1 quart	vinegar	1 liter

Put all of the vegetables together in a large bowl. Dissolve all but 1 tablespoon [15 ml.] of the salt in 2 quarts [2 liters] of water, and pour this brine over the vegetables. Let them steep for 24 hours. At the end of this time, drain off the brine and cover the vegetables with fresh water. Let the vegetables stand for about two hours. Drain off the water.

Mix together the dry ingredients, including the remaining tablespoon of salt. Mix the vinegar with 2½ cups [625 ml.] of water, bring to a boil, and gradually stir into it the dry ingredients. Bring this mixture to a boil. Pour the hot vinegar mixture over the vegetables. Fill jars with the hot mixture, cover and process *(pages 30-31)*.

WOMAN'S INSTITUTE LIBRARY OF COOKERY

Preserved Fruits and Vegetables

Candied Fruit Peel

To make about ¼ pound [125 g.]

1 lb.	halved grapefruit, orange or lemon shells, all fruit removed	½ kg.
about 9 cups	sugar	about 2¼ liters
2 quarts	water	2 liters

Scrape any remaining fruit pulp out of the grapefruit, orange or lemon shells. Cut the shells into quarters or eighths. Cover with water and bring to a boil to take away the bitter flavor. Drain, cover with fresh water and cook the peel until tender—about one hour. Drain well. Make a heavy syrup by boiling 8 cups [2 liters] of the sugar and the water together until the syrup reaches 222° F. [106° C.] on a candy thermometer and spins a thread when a little is dropped from a spoon. Cook the peel in the syrup until it is transparent and glossy—about two to three hours. Drain the peel and let it dry on a rack before rolling it in the remaining sugar.

THE WOMAN'S AUXILIARY OF OLIVET EPISCOPAL CHURCH
VIRGINIA COOKERY—PAST AND PRESENT

Citrons in Syrup

Cedri Sciroppati

A citron is an oval fruit—yellow when ripe—that looks like a large lemon but has a very thick, often bumpy, skin. It is grown in California and Florida, where the crop is picked from November through January. Citrons are obtainable at fine fruit stores.

To make about 3 pints [1 ½ liters]

2 lb.	green citrons	1 kg.
4 cups	sugar	1 liter
2½ cups	water	625 ml.

Place the citrons whole in cold water for two to three days, changing the water every day. Then boil them slowly in fresh water for three to four hours, until soft to the touch and transparent. Let them cool. Cut each citron in half and remove the soft part of the fruit with a teaspoon; discard this pulp. Put the halves of thick skin back in cold water for another two days, changing the water daily.

Cook 1 cup [¼ liter] of the sugar with 1 cup of the water over low heat, stirring until the sugar dissolves. Then boil without stirring for five minutes to make a syrup. Immerse the citrons in the syrup and boil for five minutes. Let the citrons soak in the syrup for seven days.

Cook the remaining sugar and water together to dissolve the sugar, then boil for five minutes to make another batch of syrup. Drain the first batch of syrup from the citrons, add the new syrup to the old and boil them together for about 10 minutes. Pour all of the syrup over the citrons. Let them stand for five to six days.

Remove the citrons and boil the syrup for 10 minutes. Pour it back over the citrons. Repeat three or four times on subsequent days, to eliminate all of the moisture from the citrons. They will become shiny and transparent, like candied fruit, and the syrup will become as thick as light cream. Put the citrons in jars and cover them with their syrup. Seal.

ERINA GAVOTTI (EDITOR)
MILLERICETTE

Preserved or Candied Ginger

Make sure you buy fresh ginger root that does not have fibers. It is devastating to go through the considerable processing this recipe takes and then find that the ginger slices are too fibrous to eat. The best ginger I have ever had was the kind that is tipped with red.

Wash the ginger; with a small, sharp paring knife, scrape off the outer covering. Slice enough of the ginger crosswise into thin (¼- to ⅜-inch [6- to 9-mm.]) slices to make 1 pound [½ kg.]. If you have no scale, use a generous quart [1 liter] of the ginger slices.

To make about 5 cups [1 ¼ liters]

about 1½ lb.	young, tender fresh ginger	about ¾ kg.
2 quarts	water	2 liters
3 or 4 cups	sugar	¾ or 1 liter
1 cup	light corn syrup	¼ liter
1	lemon, sliced and seeded	1

Place the ginger in a wide, large (at least 4-quart [4-liter]) saucepan. Add the water. Bring to a boil, cover, and boil gently until the ginger is just tender when tested with the tip of a paring knife—about 20 minutes. Add 1 cup [¼ liter] of the sugar, stirring until the sugar is dissolved and the mixture returns to a boil. Cover, and let stand off heat, at room temperature, for several hours or overnight.

Bring to a gentle boil; boil gently, covered, for 15 minutes. Add the corn syrup and lemon, and continue boiling gently for 15 minutes. Remove the cover and continue boiling gently for 15 minutes longer. Stir occasionally during these 15-minute cooking periods. Cover, and let stand off heat, at room temperature, for several hours or overnight.

Throughout the next cooking periods, stir from time to time to prevent scorching and to distribute the ginger slices in the syrup. Bring to a boil; stir in 1 cup of the sugar, boil gently, uncovered, for 30 minutes. Stir in the remaining 1 cup of sugar and cook until the mixture returns to a boil. Cover, and let stand off heat, at room temperature, for several hours or overnight.

Cook briskly, uncovered, to reduce the syrup until it reaches 222° F. [106° C.] on a candy thermometer, or until the syrup—allowed to drip from the side of a wooden spoon—appears to hang from the spoon and the entire surface of the mixture is covered with bubbles that break uniformly. When you are finished cooking them, the ginger slices should look translucent.

For preserved ginger, pour the slices while boiling hot into clean, hot 1-cup widemouthed jars.

For candied ginger, drain the slices after they have finished cooking, reserving the syrup. Allow the slices to dry, uncovered, overnight on a wire rack set over wax paper. When dry, roll the slices in 1 cup of sugar; then allow them to stand, uncovered, overnight on a wire rack to dry further. Store in a tightly covered container. Strain the reserved syrup for use over ice cream or fruit or cake.

CECILY BROWNSTONE
CECILY BROWNSTONE'S ASSOCIATED PRESS COOK BOOK

Preserved Green Figs

Confiture de Figues

The figs are partly peeled to allow the syrup to penetrate the fruit. Be careful not to cut through the white inner casing.

To make 3 pints [1 ½ liters]		
3 lb.	slightly underripe green figs (about 5 cups [1 ¼ liters])	1 ½ kg.
4 cups	sugar	1 liter
2 cups	water	½ liter
½	vanilla bean, split lengthwise	½

Carefully peel three or four narrow strips of skin from each fig. Put the sugar, water and vanilla bean into a large pan, and bring to a boil. Add the figs a few at a time, allowing the syrup to return to a boil before adding the next batch. Cook, uncovered, over the lowest possible heat for five hours, until the figs are soft and translucent. Remove the vanilla bean. Put the figs into jars and cover them with the syrup. Put on lids and process *(pages 30-31)*.

RAYMOND ARMISEN AND ANDRÉ MARTIN
LES RECETTES DE LA TABLE NIÇOISE

Preserved Ripe Figs

Confiture de Figues

To make about 5 cups [1 ¼ liters]		
2 lb.	firm ripe figs, each pricked in several places with a skewer (about 4 cups [1 liter])	1 kg.
3 ½ cups	sugar	875 ml.
2 cups	water	½ liter

Bring the sugar and water to a boil in an enameled, tinned or stainless-steel pan, and cook this syrup to a temperature of 222° F. [106° C.] on a candy thermometer.

Meanwhile, blanch the figs by dropping them into a large pan of boiling water. After one minute, drain them. Put the figs in the syrup, and cook over very low heat for about one hour, or until the temperature again reaches 222° F. The figs should remain intact and the syrup should become amber colored. Put into jars, cover and process *(pages 30-31)*.

ÉLIANE THIBAUT COMELADE
LA CUISINE CATALANE

Kumquat Preserves

To make 4 pints [2 liters]		
2 lb.	kumquats (about 2 quarts [2 liters])	1 kg.
1 ½ tbsp.	baking soda	22 ml.
4 cups	sugar	1 liter

Wash the kumquats and sprinkle them with the baking soda. Cover with 4 cups [1 liter] of boiling water and rinse the fruit thoroughly two or three times under cold water. Drain well, and prick each kumquat to prevent it from bursting. In an enameled, tinned or stainless-steel pan, bring to a boil enough water to cover the fruit. Drop in the kumquats. Cook for about 15 minutes, or until tender. Remove the kumquats with a slotted spoon, draining them well, and set them aside. Add the sugar to the water remaining in the pan, and boil together for about 10 minutes. Add the drained kumquats to the boiling syrup, and cook until the fruit is clear and transparent. Remove the pan from the heat and, using a slotted spoon, carefully place the fruit on shallow trays. Pour the syrup over the fruit and let stand overnight to plump the kumquats. The next morning, reheat the kumquats in their syrup. Ladle into hot jars and process *(pages 30-31)*.

JACQUELINE WEJMAN
JAMS & JELLIES

Canned Peaches

To make 2 quarts [2 liters]

12	peaches, peeled, halved and pitted, 2 pits broken and the kernels reserved	12
3 cups	water	¾ liter
½ cup	sugar	125 ml.
½ cup	honey	125 ml.

Prepare syrup by boiling the water and sugar for five minutes. Add the honey. Cook the peaches and the two reserved kernels in the syrup for five to 10 minutes, or until the peaches are tender. Arrange the peaches cut sides down in quart [liter] jars. Fill the jars with the hot syrup and remove air bubbles by running a rubber spatula around the inside edge of the jars. Cover and process *(pages 30-31)*.

AMERICAN HONEY INSTITUTE
OLD FAVORITE HONEY RECIPES

Pineapple Preserved

To make about 1 ½ pints [¾ liter]

1	large ripe pineapple, peeled, quartered, cored and sliced, rind and core reserved	1
	sugar	

Measure enough sugar to equal the volume of the prepared pineapple; measure 1¼ cups [300 ml.] of water to each 2 cups [½ liter] of sugar.

Boil the water, sugar, and pineapple rind and core together for 15 minutes. Strain the syrup and put it in an enameled, tinned or stainless-steel pan with the pineapple slices. Bring to a boil, remove from the heat and let the slices stand for one day in the syrup.

Next day, strain off the syrup again, boil it, and pour again over the sliced pineapple and allow it to stand for another day. Repeat this process the third day. It is then complete. Put the pineapple into jars with the syrup poured over. Refrigerate or process *(pages 30-31)*.

CATHERINE FRANCES FRERE (EDITOR)
THE COOKERY BOOK OF LADY CLARK OF TILLYPRONIE

Rhubarb (or Pie Plant)

To make 8 pints [4 liters]

10 lb.	rhubarb, leaves removed, stalks trimmed and cut into ½-inch [1-cm.] pieces (about 4 quarts [4 liters])	4½ kg.
	sugar	

Put the rhubarb in an enameled, tinned or stainless-steel pan, mixing in ½ cup [125 ml.] of sugar for each quart [1 liter] of raw fruit. Cover, and let the mixture stand for about four hours to draw out the juice. Bring the mixture slowly to a boil; let it boil no more than one minute, or the pieces will break up. Alternatively, bake the sugared rhubarb in a heavy, covered pan in a 275° F. [140° C.] oven for one hour. Put into jars and process *(pages 30-31)*.

RUTH HERTZBERG, BEATRICE VAUGHAN AND JANET GREEN
PUTTING FOOD BY

Red Pears

Rote Birnen

Choose pears that are juicy, but not too soft. They may be cooked in exactly the same way with raspberry juice.

To make about 5 pints [2 ½ liters]

3 lb.	pears, peeled, cored and halved or quartered (about 2½ quarts [2½ liters])	1½ kg.
3 lb.	cranberries (about 4 quarts [4 liters])	1½ kg.
2 cups	sugar	½ liter
1	cinnamon stick	1

Simmer the cranberries gently for about 15 to 20 minutes, until they are pulpy. Pour into a jelly bag and let the juice drip through. There should be about 4 cups [1 liter] of juice.

Put the juice into a pan with the pears, sugar and cinnamon. Skimming occasionally, cook over low heat until the pears are soft and the juice has reached the jelling point, 10 to 20 minutes. If the pears are cooked before the juice is ready, transfer them to jars with a slotted spoon and continue to cook the juice to the desired stage. Pour the juice over the pears in the jars. Cover and process *(pages 30-31)*.

SOPHIE WILHELMINE SCHEIBLER
ALLGEMEINES DEUTSCHES KOCHBUCH FÜR ALLE STÄNDE

Brandied Cherries

Ciliege sotto Spirito

To make about 3 pints [1 ½ liters]

1½ lb.	cherries, stems removed (about 1½ quarts [1½ liters])	¾ kg.
⅔ cup	sugar	150 ml.
4 cups	brandy	1 liter

Place the cherries, sugar and brandy in a jar and seal. Shake the jar about once a week. Let the cherries stand for four to five months before using.

ERINA GAVOTTI (EDITOR)
MILLERICETTE

Spiced Cherries in Brandy

Ciliege

Firm-fleshed cherries—preferably sour cherries—should be used for this recipe.

	To make 4 quarts [4 liters]	
4 lb.	cherries, half of each stem cut off, each cherry pricked in several places with a needle (about 4 quarts [4 liters])	2 kg.
about 6 cups	brandy or *eau de vie*	about 1½ liters
about 1 cup	sugar	about ¼ liter
1 tsp.	whole cloves	5 ml.
2	cinnamon sticks, broken into pieces	2
¼ tsp.	coriander seeds	1 ml.
2	blades mace	2
1 tsp.	black peppercorns	5 ml.

Place the cherries in a large jar or crock. Pour in enough brandy or *eau de vie* to cover them, and add ⅔ cup [150 ml.] of sugar for each 4 cups [1 liter] of brandy used. Tie the spices in a cheesecloth bag and add. Tightly cover the jar or crock, and leave for six weeks to two months. Remove the spice bag, and again cover the jar or crock. The cherries will keep for up to one year.

IL RE DEI CUOCHI

Cherries in Eau de Vie

Cerises à l'Eau de Vie

You may add 1 cup [¼ liter] of water to the sugar to help it melt and form a syrup.

	To make 8 pints [4 liters]	
4 lb.	cherries, stems trimmed (about 4 quarts [4 liters])	2 kg.
2 quarts	*eau de vie* or brandy	2 liters
3 or 4	whole cloves	3 or 4
1	small cinnamon stick	1
2 cups	sugar	½ liter

Put the cherries in jars with the cloves and cinnamon. Melt the sugar over low heat, and cook until this syrup reaches the hard-ball stage, 250° F. [120° C.] on a candy thermometer. Cool the syrup slightly, then stir in the *eau de vie*. Mix well and let cool completely. Pour the syrup over the cherries. Seal the jars.

VIARD AND FOURET
LE CUISINIER ROYAL

Preserved Greengage Plums

Prunes à l'Eau de Vie

Instead of preserving these plums in jars with their syrup, you may drain them after their final two days in the syrup, and dry them in the sun or in a 250° F. [120° C.] oven until they are sugary, about three hours. Pack them in boxes with sheets of wax paper between the layers.

	To make about 5 pints [2½ liters]	
3 lb.	slightly underripe greengage plums, half of each stem cut off, each plum pricked in several places with a needle (about 2½ quarts [2½ liters])	1½ kg.
½ tbsp.	salt	7 ml.
6 cups	sugar	1½ liters
½ cup	*eau de vie*	125 ml.

Place the plums in a saucepan and cover them generously with cold water. Heat slowly until the water is hot but not trembling. Remove the pan from the heat and add the salt. Leave for at least one hour.

Put the pan over low heat and warm the plums, stirring gently, until they turn green again. Heat to just under the boiling point, and cook until the plums begin to rise to the surface. As they rise, remove them with a slotted spoon and plunge them into a bowl of cold water.

Heat the sugar with 2½ cups [625 ml.] of fresh water, stirring until the sugar dissolves; boil for five minutes.

Drain the plums and pour the boiling syrup over them. Leave for 24 hours. The next day, strain off the syrup and boil it for about 10 minutes; pour it back over the plums and leave for another 24 hours. The next day, repeat the operation. Let the plums macerate in the syrup for two days.

Lift the plums out of the syrup with a slotted spoon, and place them in jars. Bring the syrup to a boil, and let it cool slightly. Stir the *eau de vie* into the syrup and pour it over the plums. Cover and process *(pages 30-31)*.

E. AURICOSTE DE LAZARQUE
CUISINE MESSINE

Bourbon Peaches

To make about 9 quarts [9 liters]

9 lb.	ripe peaches, scalded and peeled (about 9 quarts [9 liters])	4 kg.
6½ cups	bourbon whiskey	1 ¾ liters
18 cups	sugar	4 ½ liters
4 cups	water	1 liter
four 6-inch	cinnamon sticks, broken	four 15-cm.
2 tbsp.	whole cloves	30 ml.

In a large pan, dissolve the sugar in the water over low heat. Tie the cinnamon and cloves in a cheesecloth bag and add them to the pan. Bring this sugar syrup to a boil. When the syrup is clear, add the peaches, a few at a time, and simmer until they are barely tender, approximately five minutes. Do not overcook them.

Drain the peaches on a platter, returning the excess syrup to the pan, and repeat until all of the peaches are cooked. Boil the syrup until it is slightly thickened (222° F. [106° C.] on a candy thermometer). Cool slightly. Stir the bourbon into the syrup. As the peaches drain, place them in hot jars. Cover them with the bourbon syrup. Cap and process *(pages 30-31)*. Store in a cool, dark, dry place.

JEAN HEWITT
NEW YORK TIMES NEW ENGLAND HERITAGE COOKBOOK

Basic Rum-Pot Fruit Recipe

Grundrezept für Rumtopf

Traditionally, the rum-pot is begun in early summer, placed in a cool, dark cellar, and new layers of fruit are added throughout the summer. The fruit is ready to use by the beginning of Advent (early December).

As the rum-pot, you can use a 1- to 2-gallon [4- to 8-liter] stoneware crock with a lid. All of the fruit for the rum-pot should be ripe, but not overripe. Pick over the fruit, remove any damaged parts, and pit or core it before measuring it.

To make 1 to 2 gallons [4 to 8 liters]

3 cups	strawberries, hulled	¾ liter
about 2 cups each	prepared gooseberries, cherries, red currants, black currants, raspberries, apricots, plums and pears	about ½ liter each
about 10 cups	sugar	about 2½ liters
about 2 quarts	rum	about 2 liters

Sprinkle the strawberries with 1½ cups [375 ml.] of the sugar. Cover, and let stand for 30 minutes. Place the sugared strawberries in the rum-pot and pour in enough rum to cover the fruit by about two finger-widths. Cover the pot with plastic wrap, and put on the lid. Place in a cool room, and stir the fruit gently every two or three days.

Starting with the gooseberries, or as the fruits ripen, add the remaining fruits in layers; mix 1 cup [¼ liter] of fruit with ½ cup [125 ml.] of the sugar before adding it to the pot, and each time add enough rum to cover the fruit by two finger-widths.

After the first two weeks, you will only need to shake or stir the pot every two weeks. But you must always make sure that the top layer of fruit is covered by two finger-widths of rum. At the end of October or the beginning of November, add another cup of rum to the pot. By the beginning of December, the fruit will be ready to eat.

ARNE KRÜGER AND ANNETTE WOLTER
KOCHEN HEUTE

Tutti-Frutti

Often known as making a brandy pot, this is a very old method of preserving raw fruit as it comes into season. After the last addition of fruit and sugar, the jar should be left for about six weeks to let the contents mature.

To make 6 to 7 quarts [6 to 7 liters]

about 8 lb.	ripe fruit in season	about 3½ kg.
2½ cups	brandy	625 ml.
	superfine sugar	

Take a large stoneware jar and put the brandy into it. Add 2 pounds [1 kg.] of any ripe fruit that is in season, with 2 cups [½ liter] of superfine sugar. You can continue to add fruit and sugar in these proportions until the jar is full, only each time you must stir the mixture, right from the bottom, with a wooden spoon. Red currants, raspberries, strawberries and cherries must be stemmed and picked over; apples, pears, plums, bananas, oranges, pineapples and peaches must be peeled, cored and cut small. Of course, it is not necessary to add 2 pounds of the same kind of fruit—you could have a batch of mixed raspberries and strawberries, for instance. The more variety you put in, the better. The jar must be kept very closely covered.

MAY BYRON
MAY BYRON'S JAM BOOK

Brandied Fruit Mélange

Any combination of fruits may be used. It is well, however, to consider the flavor and the color of the syrup before adding the fruits. If an amber syrup is desired, use light fruits such as apricots, grapes, peaches or pears; if a dark syrup, use dark fruits such as berries, black cherries, plums and so on. Since only fresh fruits as they ripen should be used, it is wise to make a schedule of the fruits that will be on the market in your part of the country. The first fruit is universally the strawberry, which is included in the brandy base.

The Creoles have built quite a ceremony around the annual mélange, so they serve it during the Christmas holidays and during Mardi Gras. This recipe is a refinement of a mélange recipe given me in New Orleans.

To make about 4 gallons [16 liters]

about 10 quarts	prepared fresh fruits: hulled strawberries; pitted cherries; whole raspberries; peeled, pitted and sliced apricots and peaches; whole blueberries; pitted plums; and peeled, cored and sliced pears	about 10 liters
about 10 quarts	sugar	about 10 liters
about 2½ quarts	brandy	about 2½ liters

Brandy base

2 cups	brandy	½ liter
1½ quarts	firm, ripe strawberries, hulled	1½ liters
6 cups	sugar	1½ liters
2 cups	kirsch	½ liter
2 cups	sherry	½ liter
2	cinnamon sticks	2
1 tbsp.	chopped fresh ginger	15 ml.
1 tbsp.	whole cloves	15 ml.
1 tbsp.	whole allspice	15 ml.
1 tbsp.	grated lemon peel	15 ml.
1 tbsp.	grated orange peel	15 ml.

For the brandy base, crush the strawberries and simmer them in their own juice until they are tender, about five minutes. Let the berries drip in a jelly bag; discard the pulp. Bring the strawberry juice to a boil, add the sugar and stir until it dissolves. Cool the syrup. Put the brandy, kirsch, sherry, spices and grated lemon and orange peels into a 4-gallon [16-liter] crock. Add the strawberry syrup to the other ingredients in the crock. Stir; then cover the crock. Let the mixture stand for at least one week.

As fruits are available, add equal quantities of fresh fruits and sugar, stirring after each addition. Never add more than 2 quarts [2 liters] of sugared fruit at one time. Two or three kinds of fruit may be added at the same time. For each 2 quarts of sugared fruit, add 2 cups [½ liter] of brandy. More spices also may be added. Continue this process until the crock is filled. Put the cover on the crock and tie a cloth over it. Let it stand without disturbing it for two or three months.

MARION BROWN
PICKLES AND PRESERVES

André Daguin's Preserved Spiced Fruits in Red Wine with Armagnac

Pears and peaches for this recipe should be not quite ripe. The day before you plan to serve any of these fruits, put them into a bowl and stir in ⅔ cup [150 ml.] of uncooked red wine. Chill overnight. Serve half a pear or one peach or one plum per serving, with a little of the syrup. Serve the pears or peaches with an extra grinding of black pepper. If any of the fruits are too sweet, add lemon juice to taste.

To make 1 pint [½ liter]

4	large sour plums or peaches, unpeeled, each pricked twice with a needle, or 2 pears, peeled	4
1 tsp.	vinegar or fresh lemon juice (optional)	5 ml.
⅓ cup	Armagnac	75 ml.
2 cups	full-bodied red wine	½ liter
¾ cup	sugar	175 ml.
½ tsp.	black peppercorns or 2 cinnamon sticks	2 ml.

If using plums, soak them in the Armagnac for 24 hours. If using peaches, drop them into simmering water to cover, leave for five minutes, drain the peaches and peel off the skins. If using pears, drop them into water acidulated with the vinegar or lemon juice.

In an enameled, tinned or stainless-steel pan, heat the wine and sugar, stirring with a wooden spoon until the sugar is dissolved. Bring to a boil, and boil for one minute. Add the peppercorns and the Armagnac if using pears or peaches. Add the cinnamon sticks if using plums.

Wash a glass-topped preserving jar and a rubber ring. Keep hot. For plums or peaches, pack the hot jar with the fruit and pour in the hot syrup. For pears, drain and add them to the syrup; then simmer for two to three minutes. Pack the pears in the jar and pour in the syrup, filling the jar to within ½ inch [1 cm.] of the top.

Use a clean wooden spoon to gently stir the fruit in order to let any air bubbles escape. With a paper towel, wipe the mouth of the jar clean of any syrup. Set the rubber ring in place and close the jar at once. Gently lower it onto a rack set in a deep pot filled with enough boiling water to cover the jar by at least 1 inch [2½ cm.]. Bring to a boil, cover the pot and boil vigorously for one hour. Do not be tempted to reduce the cooking time; the wine must be fully cooked to avoid fermentation. Using tongs, transfer the jar to a rack to cool.

When the jar is completely cold, test for a full seal by loosening the clamp and holding the jar from the top. If it doesn't open, the seal has been completed. Be sure to have your other hand underneath to catch the jar if necessary. If there is a bad seal, do not store the fruits. Instead, refrigerate and use within the following week. Label a well-sealed jar and store in a cool place for three months before opening.

THE PLEASURES OF COOKING

Apple Rings

Apfelringe

Among the apple varieties suitable for drying are Baldwin, Grimes Golden, Jonathan, McIntosh, Melrose, Northern Spy, Rhode Island Greening, Russet, Spitzenburg, Stayman, Wealthy and Winesap. The peelings may also be dried; they make a tasty tea just like rose-hip tea. As with other dried fruit, soak the apple rings for a few hours before use.

To make about 2 quarts [2 liters]

9 lb.	apples (about 8 quarts [8 liters])	4 kg.

Store the well-ripened apples for a short time in a cool, airy place until they are quite mellow and soft. Then peel and core the fruit, and cut it into rings. Line the rings up on a string, and hang them in a dark, well-ventilated place. After about a week, they will be dry but still elastic. Pack them in sealed jars or plastic bags.

EIKE LINNICH
DAS GROSSE EINMACHBUCH

Naturally Dried Fruits

Fruits Secs sans Être Confits

To dry plums in the same way, gather the fruit when very ripe—preferably after the fruit has fallen from the trees. Drying time will be about four hours.

Peaches for drying should be picked from the trees when fully ripe, then halved and pitted. Drying time will be about six hours. Halfway through the drying process, flatten the peach halves so that they will dry evenly.

Dry apricots in the same way as the peaches, but remove the pits by splitting the apricots without splitting them in half. Drying time will be about four hours.

To make about 2 quarts [2 liters]

4 lb.	cherries, stems left on (about 4 quarts [4 liters])	2 kg.

Arrange the cherries in a single layer on wire racks. Put them into a preheated 225° F. [110° C.] oven. Turn off the heat and leave the cherries in until the oven is cold.

Remove the cherries, turn them over on the racks, reheat the oven to the same temperature, and return the cherries to the oven until they are sufficiently dried, about two hours. Let cool, then tie them in bunches and store in a dry place.

MENON
LA CUISINIÈRE BOURGEOISE

Dried Pears

Pera Secche

To make about 3 pounds [1 ½ kg.]

20 lb.	cooking pears, peeled, stems trimmed and lightly scraped (about 12 quarts [12 liters])	9 kg.
2 cups	sugar	½ liter
5 cups	water	1 ¼ liters

Place the pears in a large saucepan of cold water, bring to the boiling point, and cook over medium heat for about 30 minutes, or until the pears begin to soften. Remove them with a skimmer, plunge into cold water and drain.

Simmer the sugar and water together for five minutes; let cool. Immerse the pears in this syrup and set them aside for two hours.

Drain the pears, reserving the syrup. Arrange the pears, stems upward, on wire racks. Leave them overnight in a 200° F. [100° C.] oven. The next morning, dip the pears again in the syrup, then put them back in the oven. Continue this operation for four days. The last time, do not take the pears out of the oven until they are completely dry to the touch. The pears will keep for a long time if they are stored in airtight jars or bags.

IL CUOCO PIEMONTESE RIDOTTO ALL'ULTIMO GUSTO

Prune Leather

To make about ½ pound [¼ kg.]

6 lb.	pitted plums (about 4 quarts [4 liters])	2¾ kg.
1½ cups	water	375 ml.
3 tsp.	almond extract	15 ml.
	honey	

Place the plums and water in a large enameled, tinned or stainless-steel pan and cook over low heat until the fruit is tender, about 20 minutes. Drain off the juice and save it for a breakfast drink. Purée the pulp through a food mill or sieve. Discard the skins; they are too tough to be used in the leather. Add the almond extract and honey to taste. Then spread the pulp about ¼ inch [6 mm.] thick on oiled baking sheets, or on baking sheets lined with freezer paper or plastic wrap.

Cover each baking sheet with a single layer of cheesecloth or plain brown paper to keep out dust and insects, and place the sheets in a warm, dry place to dry. Depending upon the weather, the pulp will dry in one to two weeks. Drying time can be shortened to six to 24 hours by placing the baking sheets in a 120° to 150° F. [48° to 65° C.] oven, with the oven door left slightly ajar to allow moisture to escape.

When the leather is dry enough to be lifted or gently pulled from the baking sheets, put it on wire racks so that it can dry on both sides. Dust the leather with cornstarch or arrowroot when all of the stickiness has disappeared, then stack the leather in layers with freezer paper, wax paper or aluminum foil between each sheet. Cover the stack with more paper or foil, and store it in a cool, dry place.

CAROL HUPPING STONER (EDITOR)
STOCKING UP

Peach Leather

Apple and quince leathers are made in the same fashion, only a little flavoring of spice or lemon is added to them.

To make about ¼ pound [125 g.]

1 lb.	peaches, peeled, halved, pitted and cut into pieces	½ kg.
½ to 1 cup	sugar	125 to 250 ml.

Add ½ cup [125 ml.] of the sugar to the peaches. Cook gently, stirring and mashing the fruit, until the peaches are thick and dry—about 15 minutes. Cool the peach mixture, then spread it in thin sheets on greased boards or large platters, and set it in the sun. When well dried, sprinkle the peach leather with additional sugar, roll up the sheets and store them in thick paper bags. Peach leather will keep perfectly from season to season.

THE WOMAN'S AUXILIARY OF OLIVET EPISCOPAL CHURCH
VIRGINIA COOKERY—PAST AND PRESENT

Tangerine Peel

Naartje Peel

Every Malay housewife worthy of the name has on her pantry shelf a tin or bottle containing dried *naartje peel,* which she uses to flavor cakes and puddings.

To make about 1 cup [¼ liter]

3 lb.	tangerines, peels only	1½ kg.

Dry the peel in the sun or in an oven preheated to 200° F. [100° C.]. When it is thoroughly dry and hard, put the peel into a jar to store it. Pound it to a powder just before using.

HILDA GERBER
TRADITIONAL COOKERY OF THE CAPE MALAYS

Black Olives in Brine

Olives Noires en Saumure

Green olives may be used for this recipe. Instead of being pricked with a needle, each green olive should be tapped with a mallet to crack the flesh without crushing it.

To make about 4 quarts [4 liters]

4 lb.	black olives, each pricked in 3 or 4 places with a needle	2 kg.
¾ cup	salt	175 ml.
1	bay leaf	1
2	sprigs fresh fennel	2
24	coriander seeds	24
½	orange, peel only, pared in thin strips	½

Place the olives in a large crock or jar, and cover them with water. Allow the olives to soak for 10 days, changing the water every day. In a saucepan, combine 2 quarts [2 liters] of fresh water with the salt, bay leaf, fennel, coriander and orange peel. Boil this brine for 15 minutes, then let it cool. Drain the olives. Cover them with the cold brine. Store them in a cool place for at least one week before using them. Refrigerated, the olives will keep for one to two months.

CÉLINE VENCE
ENCYCLOPÉDIE HACHETTE DE LA CUISINE RÉGIONALE

Black Olives

Olives Noires

Olives become progressively less bitter as they turn from green to violet to brownish black and finally, when very ripe, to black. Only black olives have lost enough bitterness to be prepared in this way.

To make about 5½ quarts [5½ liters]

6 to 7 lb.	very ripe olives	3 kg.
4 cups	coarse salt	1 liter
3 tbsp.	black peppercorns	45 ml.
12	garlic cloves, crushed	12
4 cups	olive oil	1 liter

Put the olives in a large bowl. Mix them with the salt, and leave for eight to 10 days, turning them with your hands every day. Be careful not to crush or bruise them. Drain the olives thoroughly, and put them into an earthenware or stoneware pot. Add the peppercorns and garlic, and pour over enough oil to cover the olives. Black olives prepared in this way keep for six months in a refrigerator, but they may be eaten after eight days.

MARIA NUNZIA FILIPPINI
LA CUISINE CORSE

Green Olives in Wood Ash

Olives Vertes à la Picholine

Refrigerated, the olives will keep safely for one or two months.

To make about 4 quarts [4 liters]

4 lb.	green olives	2 kg.
4 quarts	clean wood ash, mixed with water to make a thick, runny paste	4 liters
about 2 quarts	water	about 2 liters
¾ cup	salt	175 ml.
1	bay leaf	1
2	sprigs fennel	2
24	coriander seeds	24
½	orange, the peel only, cut into thin strips	½

In a large bowl or crock, combine the olives with the wood-ash mixture. Leave them for 10 to 12 days, stirring a few times every day, until the flesh of one of the olives is easily detached from the pit.

Rinse the olives thoroughly, then cover them with cold water and allow them to stand for 10 days, changing the water each day.

Combine the water, salt, bay leaf, fennel, coriander and orange peel. Bring this brine to a boil, boil for 15 minutes and let cool. Drain the olives, return them to the bowl or crock, and cover them with the cooled brine. Store for at least one week before using.

CÉLINE VENCE
ENCYCLOPÉDIE HACHETTE DE LA CUISINE RÉGIONALE

Bottled Tomatoes

At the height of the season, ripe tomatoes have an intensity of flavor that is matched only by their abundance. Bottled in a sauce made largely from the tomatoes themselves, they make a valuable preserve to be savored in winter when the tomato harvest is a memory.

To make about 10 pints [5 liters]

14 lb.	ripe tomatoes, 4 lb. [2 kg.] coarsely chopped	6½ kg.
2	onions, chopped	2
1	bouquet garni of thyme, oregano and 2 bay leaves	1
6 tbsp.	strained fresh lemon juice	90 ml.
	salt	
1 cup	fresh basil leaves	¼ liter

Place the chopped tomatoes in a stainless-steel, enameled or tinned pan. Set over low heat and add the onions, bouquet garni and lemon juice. Season with salt. Stir the ingredients to blend them. Stirring occasionally, let the mixture cook for about 15 to 20 minutes.

While the sauce mixture cooks, cut the cores from the stem ends of the remaining tomatoes. Dip the tomatoes—a few at a time—into boiling water for several seconds to loosen their skins, then peel, halve and seed the tomatoes. Add the seed clusters to the sauce. Pack the halved and seeded tomatoes, with a bouquet of fresh basil leaves in their midst, into preserving jars.

Strain the sauce to remove skins and seeds. Fill each jar to within ½ inch [1 cm.] of the top with sauce. Tap the jars on the table to remove air bubbles. Cover the jars, and process *(pages 30-31)*. Label and store.

PETITS PROPOS CULINAIRES III

Ma Comp's Soup Seasoning

The seasoning is used generously (two or three spoonfuls in a bouillon cup) with a very hot clear meat broth such as consommé poured over it. It is also good used as a condiment, or as seasoning for a meat dish.

To make about 14 pints [7 liters]

13 lb.	ripe tomatoes, peeled, seeded and cut into chunks (about 6 quarts [6 liters])	6 kg.
5 lb.	okra, trimmed and sliced into rounds (about 2½ quarts [2½ liters])	2½ kg.
8	large onions, chopped	8
1 lb.	brown sugar	½ kg.
2 tbsp.	freshly ground black pepper	30 ml.
1 cup	salt	¼ liter
1 tbsp.	ground allspice	15 ml.
1 tbsp.	ground cloves	15 ml.
½ cup	celery seed	125 ml.

In an enameled, tinned or stainless-steel pan, combine the vegetables with the sugar and seasonings. Stirring frequently, cook very slowly until the mixture is of marmalade consistency. On an old-fashioned stove, the mixture sat at the back of the stove all day; on a modern stove, it will cook perfectly in six to 12 hours, depending on the heat. Put in jars and cover. Process *(pages 30-31)*.

THE HAMMOND-HARWOOD HOUSE ASSOCIATION
MARYLAND'S WAY

Sweet Green Tomatoes

Grüne Zuckertomaten

To make about 5 pints [2 ½ liters]

5 lb.	green tomatoes (about 2½ quarts [2½ liters])	2½ kg.
	sugar	
2	lemons, the peel only, pared into thin strips and scalded	2
1 tsp.	vanilla extract	5 ml.

Place the tomatoes in a saucepan, pour in enough boiling water to cover and simmer until soft. Drain the tomatoes, reserving the water. Measure the tomatoes and measure 3 cups [¾ liter] of sugar for each 4 cups [1 liter] of tomatoes. Stir the sugar into the cooking water, and boil for about five minutes, or until a clear syrup forms. Pour the syrup over the tomatoes and let them stand for 12 hours.

Boil the tomatoes in the syrup for 20 minutes, and let the mixture cool. Cover, and let macerate for two days.

Pour off the syrup and boil until it is thick, about five minutes. Add the lemon peel and vanilla. Pack the tomatoes into jars and pour on the syrup. Process *(pages 30-31)*.

DOROTHEE V. HELLERMANN
DAS KOCHBUCH AUS HAMBURG

Pickled Vegetables

Kim Chi

Refrigerated, these vegetables will keep for several weeks.

To make about 1 quart [1 liter]

1 lb.	Chinese cabbage, cut into 2-inch [5-cm.] pieces (about 4 cups [1 liter])	½ kg.
6 tbsp.	salt	90 ml.
4	scallions, finely chopped	4
1 tsp.	finely chopped fresh ginger	5 ml.
1	garlic clove, crushed to a paste	1
1 tbsp.	sugar	15 ml.
1 tbsp.	paprika	15 ml.
	cayenne pepper	
2 cups	water	½ liter

Sprinkle the cabbage with 5 tablespoons [75 ml.] of the salt. Let stand for two hours. Rinse off the salt and drain. Add the scallions, ginger, garlic, sugar, paprika, the remaining salt and a dash of cayenne pepper, and mix thoroughly. Pack the vegetables in a jar and pour in the water. Cover tightly. Refrigerate the vegetables for 24 hours before serving.

ALICE MILLER MITCHELL
ORIENTAL COOKBOOK

Basil Preserved in Oil

Basilic Confit dans l'Huile

Basil preserved in this way may be used wherever you would be using salt and oil as well as basil: in salads, for instance, or sweated with the vegetables for soups and stews.

To make about 1 cup [¼ liter]

2 cups	basil leaves, stemmed, washed and spread to dry on a towel	½ liter
2 tbsp.	salt	30 ml.
½ cup	olive oil	125 ml.

Put a layer of basil leaves in a jar, sprinkle with salt, and continue to make layers of basil and salt until the jar is three quarters full. Fill the jar with olive oil, then seal it and store in a cupboard. The basil will be ready to use after a month.

TINA CECCHINI
LES CONSERVES DE FRUITS ET LÉGUMES

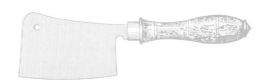

Corn on the Cob in Brine

To make about 2 quarts [2 liters]

5 lb.	very small, young, tender ears of corn	2½ kg.
1 quart	water	1 liter
2 tsp.	salt	10 ml.
½ cup	sugar	125 ml.
2 or 3	bay leaves	2 or 3
1 tsp.	black peppercorns	5 ml.
1	lemon, the peel only, cut into large pieces	1
1½ tbsp.	fresh lemon juice	22 ml.

Trim the length of the ears to fit the corn into large jars. Drop the corn into salted, boiling water to cover, and boil for three or four minutes. Drain, and spread the ears out on a clean cloth in an airy place to dry for two hours, turning them over after one hour. Put the corn in large jars.

Boil the 1 quart [1 liter] of water, the salt, sugar, bay leaves, peppercorns and lemon peel for two minutes. Stir in the lemon juice. Strain the mixture. Cool, then pour it over the ears of corn. Cover, and process by the pressure-canner method *(pages 30-31)*. Store in a cool, dark, dry place.

To serve, put the ears in boiling water for several minutes, let them dry, then brown them in butter.

ANGELO SORZIO
THE ART OF HOME CANNING

Grape Leaves

To make 2 pints [1 liter]

1 quart	whole grape leaves, packed tightly	1 liter
2 tsp.	salt	10 ml.
1 quart	water	1 liter
1 cup	lemon juice or 2½ tsp. [12 ml.] citric acid	¼ liter

Add the salt to the water and bring to a boil. Add the grape leaves. After 30 seconds, drain the leaves in a colander set over a deep pot to reserve the cooking liquid. Stack the leaves in bunches of 10 or so. Shape each stack into a loose roll and pack the rolls vertically in pint jars.

Add the lemon juice or citric acid to the cooking liquid. Bring to a boil, then pour the mixture into the jars to cover the rolled leaves. Close the jars. Process for 15 minutes in a boiling water bath.

UNIVERSITY OF CALIFORNIA COOPERATIVE EXTENSION
SAFE METHODS FOR PREPARING PICKLES, RELISHES AND CHUTNEYS

Potted Mushrooms

These mushrooms are not for storing for any length of time, but should be eaten within a month or so. Store the potted mushrooms in a refrigerator.

To make about 5 cups [1 ¼ liters]

1½ lb.	mushrooms	¾ kg.
1½ tbsp.	butter	22 ml.
	salt and pepper	
	cayenne pepper	
	ground mace	
1 tbsp.	anchovy paste	15 ml.
	clarified butter	

Wipe the mushrooms carefully to remove any grit. Put them into a scrupulously clean saucepan and cook slowly until the juice runs. Then increase the heat and cook, stirring occasionally, until all of the juice has disappeared. Then add the butter, seasonings and spice. Cook until the butter is absorbed, then add the anchovy paste. Cook gently for a little longer, then press into small bowls or pots. Cool, and cover with clarified butter.

ROSEMARY HUME AND MURIEL DOWNES
JAMS, PRESERVES AND PICKLES

Crock Sauerkraut

Tightly packed in well-covered containers, this sauerkraut can be safely kept in the refrigerator for six months or more.

Sauerkraut may be served cold in salads or hot with meats. The sharpness of its flavor will depend on how long it is cooked. For the most tang and greatest crispness, simply heat it. For a milder flavor, cook it longer. Late cabbage is best for sauerkraut as it is higher in sugar. Take care to measure the salt accurately—use a knife to level the tablespoon. The cabbage will not ferment properly if you add too much or too little salt.

To make about 10 gallons [40 liters]

50 lb.	firm, mature cabbages, quartered and cored, outer leaves discarded (about 10 gallons [40 liters])	22½ kg.
about 3 cups	coarse salt	about ¾ liter

With a shredder or sharp knife, shred 5 pounds [2½ kg.] of cabbage to the thickness of a dime. Place in a large mixing bowl. Sprinkle 3 tablespoons [45 ml.] of salt over the cabbage. Mix well with your hands or a stainless-steel spoon.

Wash a 10-gallon [40-liter] crock with soapy water, rinse, and scald it with boiling water. Drain thoroughly. Pack the salted cabbage, batch by batch, into the crock. Juices will form as you pack and press the cabbage down.

Repeat the shredding and salting of the cabbage until the crock is filled to within no more than 5 inches [13 cm.] of the top. Make sure the juice covers the cabbage. If not, make additional brine by mixing 1½ tablespoons [22 ml.] of salt with 4 cups [1 liter] of boiling water. Cool to room temperature before adding to the crock.

Now the cabbage needs to be covered and weighted down to keep it submerged in the brine. Fit one large plastic bag inside another to make a double bag. Fill with brine solution—1½ tablespoons salt to 4 cups water—and lay over the cabbage. The bag should fit snugly against the inside of the crock to seal the surface from exposure to air; this will prevent the growth of a yeast film or mold. The amount of brine in the bag can be adjusted to keep the cabbage submerged. Twist and tie to seal the bag. Cover the crock with plastic wrap, then with a heavy terry towel. Tie twine around the crock to hold the plastic wrap and towel in place. Do not open until fermentation time is completed.

Fermentation will begin the day following packing. How long it takes depends on the room temperature. For best-quality sauerkraut, a room temperature of 75° F. [23° C.] is ideal; fermentation will take about three weeks. At 70° F. [20° C.], allow about four weeks; at 65° F. [18° C.], about five weeks; and at 60° F. [15° C.], about six weeks. Temperatures above 75° F. will result in premature fermentation and possible spoilage. Keep track of the temperature so that you know when to check the sauerkraut. Remove the cover. Fermentation is complete if bubbling has stopped and no bubbles rise when the crock is tapped gently.

The old-fashioned way. Instead of weighting the cabbage with a brine-filled plastic bag, you can give the sauerkraut daily care as follows:

Cover the cabbage with a clean white cloth. Cover the cloth with a scalded heavy plate that fits snugly inside the crock. Fold the cloth over the plate. For a weight, fill clean glass jars with water; cap with lids and screw bands; scald the jars before setting them on the plate. Use enough weight to bring the brine 2 inches [5 cm.] above the plate—this makes daily skimming easier. Make additional brine if necessary. Cover the crock with a clean heavy terry towel, and top with plastic wrap to help prevent evaporation. Tie with twine. Each day, uncover the crock and remove yeast film or mold with a scalded stainless-steel spoon. Have a second jar weight ready and scalded to replace the one you remove. Replace the cloth and plate with clean ones. Cover the crock again with a clean terry towel. The sauerkraut may be stored in the refrigerator after fermentation is completed. Or, for longer keeping, it can be brought to a boil in a large saucepan, then canned in quart jars and processed in a boiling water bath for 20 minutes.

NELL B. NICHOLS AND KATHRYN LARSON
FARM JOURNAL'S FREEZING & CANNING COOKBOOK

Celery Salt
Selleriesalz

To make about 3 ½ cups [875 ml.]

½ lb.	celeriac or small bunch celery	¼ kg.
3 cups	coarse salt	¾ liter

Peel and finely grate the celeriac, or trim the bunch of celery and chop the ribs as fine as possible. Mix the celery with the salt, cover, and leave at room temperature for two days. Then put the mixture into a shallow ovenproof dish and dry it in a 300° F. [150° C.] oven, keeping the door slightly ajar, for approximately three to four hours. When the salt is completely dry, pound it until the crystals are small. Put the salt into jars and tightly close.

EIKE LINNICH
DAS GROSSE EINMACHBUCH

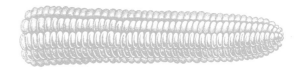

To Dry Sweet Corn

To make about 1 pint [½ liter]

24	ears of corn, kernels cut off cobs (about 1 ½ quarts [1 ½ liters] kernels)	24
½ cup	sugar	125 ml.
¼ cup	salt	50 ml.
2 tbsp.	flour	30 ml.
½ cup	cream	125 ml.

Combine the sugar, salt and flour, and mix thoroughly. Add the dry ingredients to the corn kernels and blend together. Add the cream and mix thoroughly. Spread the corn mixture in thin layers in shallow baking pans, and place to dry in a 250° F. [120° C.] oven, stirring frequently, for two to three hours. When the corn is thoroughly dried, store it in tightly covered containers.

MARY EMMA SHOWALTER
MENNONITE COMMUNITY COOKBOOK

Tomato Figs

Sun-drying these tomato figs may require three or four days. Alternatively, they can be dried in a dehydrator or a 140° F. [60° C.] oven for six to eight hours.

Tomatoes thus put up will keep indefinitely and are nearly equal to figs. Peaches may be kept in the same way.

To make about 2 ½ quarts [2 ½ liters]

8 lb.	plum tomatoes or other small tomatoes, scalded and peeled (about 3 ½ quarts [3 ½ liters])	3 ½ kg.
6 cups	brown sugar	1 ½ liters
4 cups	confectioners' sugar	1 liter

Cook the tomatoes and brown sugar together, without water, in an enameled, tinned or stainless-steel pan over very low heat until the sugar penetrates the tomatoes and they appear clear, about 30 minutes. Take the tomatoes out of the pan and arrange them on dishes in a single layer. Dry in the sun, sprinkling on a little of the cooking syrup while drying, until they are firm but not leathery. Pack in jars or boxes, with confectioners' sugar between layers.

THE BUCKEYE COOKBOOK

Dried Tomato Paste

Italian plum tomatoes are best, but other types can be used.

To make about 1 to 2 pounds [½ to 1 kg.]

20 lb.	tomatoes, quartered	9 kg.
5	large onions, chopped	5
10	celery ribs with leaves, chopped	10
10	garlic cloves, finely chopped	10
1	bunch fresh parsley, chopped	1
10	sprigs oregano, chopped	10
10	sprigs thyme, chopped	10
2	bay leaves	2
50	black peppercorns, crushed (about ½ tbsp. [7 ml.])	50
20	whole cloves	20
3 tbsp.	salt	45 ml.

Simmer everything together (do not add water—just mash the tomatoes a little so they make their own juice) for an hour or so, stirring it every time you pass the stove. Now, 2 to 3 cups [½ to ¾ liter] at a time, purée the sauce in the blender, and then put it through a sieve if you wish.

Put the pot of puréed pulp back on the stove, but turn the heat to low so that it does not scorch. Stirring occasionally, simmer very slowly, uncovered, until the pulp is reduced by half and quite thick. This will take several hours.

Next, spread the pulp ½ inch [1 cm.] thick on plates or stainless-steel baking sheets and put out in the sun. As it starts to dry, cut through the paste in a crisscross pattern, to allow air to penetrate as much as possible. Protect the paste from insects with a storm or screen window, a piece of cheesecloth, or netting. A day or two of hot sun will dry the pulp to the stage when you can scrape it off the plates or baking sheets and form it into small nonsticky balls.

If you do not live in a sunny climate, you can dry the paste in a 140° F. [60° C.] oven; this will take six to eight hours.

After the paste is rolled into small balls, let them dry for a day more at room temperature, and then store them in a tightly lidded jar. To use, dilute with a little boiling water or stock, or add a couple of balls to a batch of minestrone or spaghetti sauce.

CAROL HUPPING STONER (EDITOR)
STOCKING UP

Mincemeat

Boiled cider, which has the density of maple syrup, is bottled commercially in Vermont and is obtainable at specialty food stores. It can be made by boiling fresh cider to 222° F. [106° C.]; 8 quarts [8 liters] of fresh cider are necessary to produce the 3 cups [¾ liter] of boiled cider required for this recipe.

To make about 9 pints [4½ liters]

2 lb.	boneless lean beef, poached in water until tender—about 2 hours	1 kg.
1 lb.	beef suet	½ kg.
3 lb.	tart apples, chopped (about 3 quarts [3 liters])	1½ kg.
3 lb.	raisins	1½ kg.
2 lb.	dried currants	1 kg.
1 tbsp.	candied citron	15 ml.
3	oranges, the peel grated and the juice squeezed	3
3	lemons, the peel grated and the juice squeezed	3
5 cups	sugar	1¼ liters
4 cups	beef stock *(recipe, page 167)*	1 liter
3 cups	boiled cider	¾ liter
2 cups	tart jelly	½ liter
1 cup	molasses	¼ liter
2 tbsp.	salt	30 ml.
2 tbsp.	ground cinnamon	30 ml.
2 tsp.	grated nutmeg	10 ml.
2 tsp.	ground mace	10 ml.
2 tsp.	ground cloves	10 ml.
1 tsp.	ground allspice	5 ml.

Put the meat through a food grinder with the suet. In a large enameled, tinned or stainless-steel pot, combine the ground meat with the remaining ingredients. Stirring frequently, bring to a boil. Then reduce the heat and simmer for two hours. Put into jars. Process *(pages 30-31)*.

RUTH GRAVES WAKEFIELD
TOLL HOUSE TRIED AND TRUE RECIPES

Our Fabulously Good Mincemeat

To make about 8 quarts [8 liters]

3 lb.	lean beef brisket or rump	1½ kg.
3 lb.	fresh beef tongue	1½ kg.
1½ lb.	suet, finely chopped	¾ kg.
2 lb.	Muscat raisins	1 kg.
2 lb.	seedless white raisins	1 kg.
2 lb.	dried currants	1 kg.
½ lb.	candied citron peel, cut into thin shreds	¼ kg.
½ lb.	candied orange peel, cut into thin shreds	¼ kg.
¼ lb.	candied lemon peel, cut into thin shreds	125 g.
½ lb.	dried figs or pitted dates, chopped (optional)	¼ kg.
2 cups	sugar	½ liter
2 cups	strawberry jam	½ liter
1 tbsp.	salt	15 ml.
2 tsp.	grated nutmeg	10 ml.
2½ tsp.	ground cinnamon	12 ml.
1 tsp.	ground allspice	5 ml.
1 tsp.	ground mace	5 ml.
½ tsp.	ground cloves	2 ml.
3¼ cups	sherry	800 ml.
about 1½ quarts	Cognac	about 1½ liters

Cover the beef and the tongue with water, and cook them over low heat until tender, about two and one half to three hours. Cool the meats, peel and trim the tongue, and put the meats through the coarse disk of a food grinder or chop them by hand. In a deep crock, combine the meats with all of the remaining ingredients, adding enough Cognac to make a rather loose mixture of the fruits and meats. Mix the ingredients thoroughly, cover the crock, and let the mincemeat stand for a month or so before using it.

Check the mincemeat each week to see if absorption necessitates adding more sherry or Cognac, or both.

JAMES BEARD
DELIGHTS AND PREJUDICES

Excellent Mincemeat

To make about 2 quarts [2 liters]

3	large lemons, the peel grated, the juice strained and the hulls reserved	3
3	large tart apples	3
1 lb.	seedless raisins	½ kg.
1 lb.	dried currants	½ kg.
1 lb.	suet, cut into fine shreds	½ kg.
5 cups	brown sugar	1¼ liters
¼ cup	candied citron peel, thinly sliced	50 ml.
¼ cup	candied orange peel, thinly sliced	50 ml.
¼ cup	candied lemon peel, thinly sliced	50 ml.
1 cup	brandy	¼ liter
2 tbsp.	orange marmalade	30 ml.

Cover the lemon hulls with water and boil them until very tender—about 45 minutes. Drain, and purée the hulls through a food mill or chop them fine. Add the lemon juice and grated lemon peel. Bake the apples in a preheated 350° F. [180° C.] oven for 30 to 45 minutes, or until tender. Remove their peels and cores, and add the apple pulp to the lemon pulp. Put in the remaining ingredients one by one, mixing everything very thoroughly together. Put the mincemeat into a stoneware jar with a close-fitting lid and store in a cool place. In a fortnight it will be ready for use.

MRS. ISABELLA BEETON
THE BOOK OF HOUSEHOLD MANAGEMENT

Pickled Pig's Feet

To make about 2 quarts [2 liters]

8	pig's feet	8
½ cup	salt	125 ml.
2 quarts	vinegar	2 liters
1	small hot red chili	1
2 tbsp.	freshly grated horseradish	30 ml.
1 tsp.	black peppercorns	5 ml.
1 tsp.	whole allspice	5 ml.

Sprinkle the pig's feet with the salt and let stand for four hours. Wash the feet well in water, place them in a large pan of hot water, and cook for about two and one half hours, or until they are tender but the meat still adheres to the bones.

Pack the feet in jars. Boil the vinegar with the seasonings, and pour into the jars to within ½ inch [1 cm.] of the tops. Process in a boiling water bath for one and a half hours.

LEONARD LOUIS LEVINSON
THE COMPLETE BOOK OF PICKLES AND RELISHES

Headcheese
Hoofdkaas

To make 2 to 2 ½ pounds [1 kg.]

½	pig's head	½
4	pig's feet	4
1	onion, chopped	1
2 tbsp.	salt	30 ml.
⅔ cup	vinegar	150 ml.
1 tsp.	freshly ground pepper	5 ml.
½	nutmeg, grated	½
4	pickled sour gherkins, coarsely chopped	4
1 tsp.	ground mace	5 ml.
	lard, melted	

Put the head and feet in a saucepan, and barely cover them with cold water. Add the onion and salt, and bring slowly to a boil. Simmer over low heat for two to three hours, or until the meat is easily detached from the bones. Remove the meat from the pan and bone it carefully. Dice or chop the meat. Strain the cooking liquid.

Combine the meat with the vinegar, pepper, nutmeg, gherkins, mace and as much of the cooking liquid as necessary to make a smooth syrupy mixture. Simmer it gently for another 15 minutes.

Rinse out stoneware pots or bowls with cold water or vinegar. Ladle in the headcheese, pressing it down well and filling the containers to just below the rims. Let cool until the liquid jells. Cover the surface with a layer of melted lard. Refrigerated, the headcheese will keep for about a month.

C. A. H. HAITSMA MULIER-VAN BEUSEKOM (EDITOR)
CULINAIRE ENCYCLOPÉDIE

Pork Rillettes
Rillettes Sarthoises

This potted meat is always on the dining table in the Sarthe region in northwestern France, as butter is in Brittany.

To make about 4 to 5 pounds [2 to 2 ½ kg.]

4 to 5 lb.	fresh lean pork belly, rind removed, cut into cubes	2 to 2½ kg.
8 tbsp.	lard or ¾ cup [175 ml.] water	120 ml.
¼ cup	coarse salt	50 ml.
½ tsp.	pepper	2 ml.
2 or 3	whole cloves	2 or 3
1	bouquet garni	1

Place the meat in a heavy pan with all of the other ingredients. If you use lard, the result will be firmer than if you use water. Cover and cook over very low heat—the rillettes should stay just under the boiling point. Stir from time to time with a wooden spoon to prevent sticking.

After five to six hours, remove the pan from the heat. Crush the meat with a large wooden spatula or spoon. No piece of meat should remain whole, and the fat and lean should be thoroughly mixed.

Return the pan to the heat for about 15 minutes; when the mixture reaches the boiling point, ladle it into china or earthenware pots. There should be a layer of fat covering the surface of the meat. Allow to cool. Cover the pots with foil and refrigerate them. Use the rillettes within two months.

ÉDOUARD NIGNON (EDITOR)
LE LIVRE DE CUISINE DE L'OUEST-ÉCLAIR

Potted Rabbit

This light-textured form of potted meat will keep for up to two months in a refrigerator. Serve it as an hors d'oeuvre, accompanied by good bread and coarse salt, or as a snack.

To make about 2 pounds [1 kg.]

2 lb.	rabbit, cut into pieces	1 kg.
1½ lb.	fresh pork belly, cut into 1½-inch [4-cm.] cubes	¾ kg.
¾ lb.	hard pork fatback, cut into 1½-inch [4-cm.] cubes	350 g.
1	sprig thyme	1
½ tsp.	grated nutmeg	2 ml.
½ tsp.	ground cinnamon	2 ml.
	salt and black pepper	
8 tbsp.	lard, melted	120 ml.

Put the rabbit, pork belly and fatback into a heavy pan with the thyme and a ladleful of water. Cover, and cook over low heat or in a 300° F. [150° C.] oven for four hours, or until the meat falls off the bones.

Pour the contents of the pan into a large strainer set over a bowl. Reserve the liquid; remove and discard the bones, pork skin and thyme sprig. In a mortar or heavy bowl, crush the meat with a pestle, then tear it into shreds with two forks; this demands a little patience. One shortcut is to shred the meat in a blender or processor, but the meat must not be pulverized to sludge. (Whatever you do, don't grind it.) The final result should be unctuous and thready.

Season the meat generously with spices and pepper, and add some of the strained fat and juices, but be discreet or the rillettes will congeal too solidly as they cool. Reheat the rillettes to the boiling point, add salt to taste, and pack into jars or stoneware mugs. Cool. Cover with ½ inch [1 cm.] of melted lard, then with lids or plastic wrap. Refrigerate.

JANE GRIGSON
GOOD THINGS

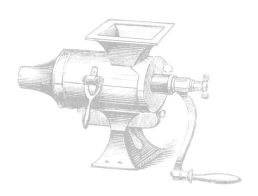

Potted Tongue

To cook a fresh tongue, first soak it in salted water for one hour. Drain, then put the tongue in a pan and cover it with fresh water. Bring to a boil, skim, and add an onion, carrot, celery rib, bouquet garni, a few peppercorns and 1 teaspoon [5 ml.] of salt. Simmer, partly covered, for about 45 minutes per pound [½ kg.], or until the tongue is tender. When the tongue is cool enough to handle, skin and trim it, then return it to the cooking liquid to cool completely.

Potted meats can be served instead of pâté at the beginning of a meal, or they make delicious sandwich fillings. This can be kept in a refrigerator for up to three weeks.

	To make about 2 cups [½ liter]	
½ lb.	cooked beef tongue, half of it finely ground, half cut into very small dice	¼ kg.
1	small garlic clove, chopped	1
	grated nutmeg	
	ground ginger	
	ground mace	
	crumbled dried thyme	
12 tbsp.	butter	180 ml.
	salt and freshly ground black pepper	

In a mortar, combine the chopped garlic clove with a pinch each of nutmeg, ginger, mace and thyme, and pound with a pestle until the thyme is almost powdered and the garlic amalgamated with the other ingredients.

Combine 8 tablespoons [120 ml.] of the butter with the ground tongue and the spice mixture, and mix to a smooth paste. Mix the diced tongue with the paste, seasoning with salt and pepper. Pack into small jars.

Clarify the remaining butter by bringing it to a boil and straining it through a piece of cheesecloth. Pour the hot butter over the potted tongue and leave to cool. Refrigerate.

MARIKA HANBURY TENISON
RECIPES FROM A COUNTRY KITCHEN

Preserved, Stuffed Goose Neck

Cou d'Oie Farci et Confit

Quatre épices, or Parisian spice, is a mixture of ground spices that may vary according to taste. A standard mixture is 1 teaspoon [5 ml.] each of cloves, nutmeg and ginger, and 1 tablespoon [15 ml.] of black pepper.

	To make 1 ½ lb. [¾ kg.]	
1	neck of a well-fattened goose, with the adjoining breast and back skin attached	1
¼ lb.	fresh mushrooms, diced	125 g.
1	shallot, chopped	1
2 tbsp.	strained fresh lemon juice	30 ml.
about ½ lb.	goose or pork fat, diced	about ¼ kg.
2	chicken livers	2
5 oz.	boneless pork loin	150 g.
2 tbsp.	Cognac	30 ml.
1	egg	1
	salt and pepper	
	quatre épices	

Put the mushrooms in a saucepan with the shallot, lemon juice and half of the fat, and place over low heat. Cover and cook for 15 minutes. Let cool.

Taking care not to split it, roll up the skin of the goose neck on itself until it is completely detached from the bone. With a small knife, remove all of the meat from the neck bones. Using the medium disk, grind the neck meat, chicken livers and pork in a food grinder. Combine the ground meats, the mushroom mixture, the Cognac and the egg, and season with salt, pepper and a pinch of *quatre épices*. Knead the mixture thoroughly. Unroll the neck skin and fill it with the mixture. Sew up both ends of the skin casing.

Pour water into a saucepan to a depth of ½ inch [1 cm.], and add the remaining fat. Place over low heat until the fat is melted (the liquid should be sufficient to cover the goose neck). Put the goose neck into the saucepan, cover and cook for two hours.

Remove the neck from the fat. Bring the fat to a boil, then let it cool to tepid. Put a ladleful of the fat into a stoneware jar, swirling it around the sides. Let it solidify. Put in the goose neck and cover the neck with the remaining fat, poured through a strainer. Seal the jar and store in a refrigerator. The meat will keep for up to two months.

CÉLINE VENCE
LE GRAND LIVRE DES CONSERVES, DES CONFITURES
ET DE LA CONGÉLATION

Goose Rillettes

Rillettes de la Mère l'Oie

To make about 10 cups [2 ½ liters]

1 ½ lb.	goose, cut into small pieces	¾ kg.
1 lb.	fresh lean pork belly, diced	½ kg.
¾ cup	water	175 ml.
1	bouquet garni	1
	salt and pepper	
1	onion, finely chopped	1
1	carrot, finely chopped	1

Put all of the ingredients into a saucepan. Cook over very low heat for about eight hours. The meat should fall off the bones. Spoon off and reserve the fat. Pick out all of the bones and mash the meat with a fork. Pack the meat into glazed stoneware jars. Pour on a layer of the reserved fat to completely cover the meat, reheating the fat if it has congealed.

When cool, cover the pots with foil and refrigerate them. The rillettes will keep for a month or two, but should be used within a week after the seal of fat is broken.

ÉDOUARD NIGNON (EDITOR)
LE LIVRE DE CUISINE DE L'OUEST-ÉCLAIR

To Pott Beef

This recipe is adapted from a 17th Century cookbook.

To make about 4 pounds [2 kg.]

5 lb.	boneless beef rump or round roast	2 ½ kg.
½ cup	finely chopped fresh sage	125 ml.
¼ cup	finely chopped fresh thyme	50 ml.
2 tsp.	ground mace	10 ml.
2	nutmegs, grated	2
2 tbsp.	salt	30 ml.
1 tbsp.	pepper	15 ml.
1 ½ lb.	butter, 1 lb. [½ kg.] softened, ½ lb. [¼ kg.] melted	¾ kg.

Roast the beef in a preheated 400° F. [200° C.] oven for about one and one half hours, or until well done. Take it from the pan. When the beef is cold, pare off all the outside crust and the fat, cut the meat with a knife into small bits, and pound these in a mortar. Beat the seasonings into the meat.

In a stoneware jar, layer the meat with the softened butter until the jar is full. Press the mixture down well, cover, and bake in a preheated 350° F. [180° C.] oven for one hour. Let the beef cool. When it is cold, pour on the melted butter to seal the top. Refrigerated, the beef will keep for two weeks.

ANN BLENCOWE
THE RECEIPT BOOK OF ANN BLENCOWE

Salt Anchovies

Anchois au Sel

The techniques of preparing and salting anchovies are demonstrated on pages 82-83.

To make about 1 quart [1 liter]

3 lb.	very fresh anchovies, heads and connecting entrails removed	1 ½ kg.
5 cups	coarse salt	1 ¼ liters

Place the fish on a layer of coarse salt and spread additional salt over them. Leave the fish for up to 24 hours, or until they exude a little water. Drain the fish on paper towels.

Put a ½-inch [1-cm.] layer of salt at the bottom of a wide-mouthed jar. Pack a layer of anchovies tightly together, head to tail, on top of the salt. Cover with another layer of salt, then with a layer of anchovies, crosswise to the first layer. Repeat these layers until the jar is full. Finish with a layer of salt.

Put a weight that fits just inside the neck of the jar on top of the anchovies. Leave for about a week, or until an oily substance rises to the top. Remove this with a spoon, then close the jar and put it in the refrigerator. To use salt anchovies, rinse them under running cold water or soak them in cold water. Refrigerated, they will keep for a year or two.

MYRETTE TIANO
CONSERVES MAISON

Salted Fish Paste

Le Melet

This and a similar Niçoise preparation known as pissalat are believed to be direct descendants of the ancient Roman fish paste called garum. In Provence, this paste is made from tiny whitebait no more than 1 inch [2 ½ cm.] long.

To make about 2 ½ cups [625 ml.]

2 lb.	whitebait	1 kg.
2 cups	coarse salt	½ liter
	thyme	
	bay leaves	
	mixed dried herbs	

In a crock or small barrel, form a layer of salt, a layer of herbs and a layer of fish. Repeat these layers until the crock is full. Weight the top layer so that the fish will be kept below the level of the brine as it forms. When the salt is completely dissolved, after about a week, remove the fish from the brine. Press them through a sieve and pack the resulting purée into pots. Cover the fish paste with a little of the brine from the crock. The paste is used like anchovy paste. Covered and kept refrigerated, it will last throughout the year.

JEAN-NOEL ESCUDIER
LA VÉRITABLE CUISINE PROVENÇALE ET NIÇOISE

Gefilte Fish

Always use at least three kinds of fish, in equal parts or otherwise, to suit your taste; the three we list are traditional, but any other fresh-water fish will do.

To make about 4 pints [2 liters]

1	whitefish, filleted, head and bones reserved	1
1	carp, filleted, head and bones reserved	1
1	pike, filleted, head and bones reserved	1
1 to 3	eggs	1 to 3
	salt and pepper	
¼ tsp.	sugar	1 ml.
3 or 4	carrots, sliced	3 or 4
2 or 3	medium-sized onions, sliced	2 or 3

Grind the raw fish quite fine, alternating pieces of the three different kinds. Add one egg for each pound [½ kg.] of ground fish. Add salt, pepper and the sugar. Shape this mixture into balls about 1 inch [2½ cm.] in diameter.

Put the carrots, onions, fish heads and bones into enough water to cover them well. Bring to a boil, and boil for 20 minutes or so, replenishing the liquid with hot water if the fish trimmings do not stay covered. Add the fish balls, reduce the heat, and simmer for 30 to 40 minutes.

Discard the fish heads and bones. Pack the fish balls, the carrots and any onions that are left intact into hot jars, and fill them nearly full with the broth.

Process the jars in a pressure canner for 90 minutes at 10 pounds of pressure [70 kPa].

VERA GEWANTER & DOROTHY PARKER
HOME PRESERVING MADE EASY

Caviar

The technique of making caviar is shown on pages 84-85.

Salmon is to be preferred for making caviar (unless you can obtain sturgeon roe), but there are those who like caviar made from the eggs of cod, tuna, whitefish and even herring.

To make 1 cup [¼ liter]

1 cup	fresh fish roe, preferably salmon	¼ liter
2 cups	cold water	½ liter
½ cup	coarse salt	125 ml.

Spread a piece of coarse-mesh wire netting over a bowl. Break up the roe and rub the pieces over the mesh so that the eggs, loosened from the membrane that holds them, drop into the bowl. Be fairly gentle in this operation—you do not want to break the eggs, nor do you want to force bits of membrane through the mesh (you will just have to pick them out later). Sometimes you can use your fingers to help free the eggs. However you do it, get the eggs into the bowl.

Make a brine of the water and salt. Stir until the salt has dissolved, then add the eggs and stir gently. Let the eggs stand in the brine for 15 to 20 minutes, swirling the mixture gently once in a while and picking out any bits of the membrane that have slipped through the mesh.

Pour the eggs into a fine-meshed sieve (ideally of nylon or stainless steel) and let them drain over a bowl in the refrigerator for about an hour.

Pack the caviar into a jar, cover closely, and refrigerate for about 12 hours before serving. If the caviar seems too salty for your taste, rinse it quickly with ice-cold water and drain it. Tightly covered, the caviar will keep in the refrigerator for about a month.

HELEN WITTY AND ELIZABETH SCHNEIDER COLCHIE
BETTER THAN STORE-BOUGHT

Barrie Davidson's Sweet Pickled Herring

This is a distinctive way of pickling herring, the results of which are enjoyed by guests at the Kilchoan Hotel, Ardnamurchan, Scotland. My fellow clansman, whose hotel it is, declares this recipe to be soundly based on a Highland tradition and so popular that he has to make it frequently in the quantities here indicated.

To make about 6 pints [3 liters]

7 lb.	fresh herring fillets, skins left on, fillets cut into 1-inch [2½-cm.] squares	3 kg.
2½ quarts	white vinegar	2½ liters
six 2½-inch	cinnamon sticks	six 6-cm.
¼ cup	whole allspice	50 ml.
¼ cup	blades mace	50 ml.
¼ cup	black peppercorns	50 ml.
¼ cup	whole cloves	50 ml.
3	large onions, thinly sliced	3
1¾ cups	sugar	425 ml.

Combine all of the ingredients except the herring and boil the mixture for 15 minutes. Then add the herring pieces. Let the herring simmer for 10 to 15 minutes. They will be "setting": that is, achieving the right degree of firmness. Then let the herring cool in the liquid.

Refrigerated and covered with the cooking liquid, the herring will keep for up to one month.

ALAN DAVIDSON
NORTH ATLANTIC SEAFOOD

Pickled Salmon

Refrigerated, this salmon will keep safely for one and one half to two months.

To make about 5 pints [2 ½ liters]

10 lb.	boned salmon	4½ kg.
	salt	
1 quart	water	1 liter
2	medium-sized onions, sliced (about 1 cup [¼ liter])	2
½ cup	olive oil	125 ml.
1 quart	white vinegar	1 liter
10	bay leaves	10
1 tbsp.	ground white pepper	15 ml.
½ tbsp.	black peppercorns	7 ml.
1 tbsp.	mustard seeds	15 ml.
½ tbsp.	whole cloves	7 ml.

Cut the salmon into small serving portions, wash, drain, and dredge them with salt. Let stand for 30 minutes, then rinse the salmon and simmer in the water until done, about 10 minutes. Drain the pieces, and place them in a sterilized crock or in Mason jars.

Meanwhile, cook the onions in the oil until yellow. Add the other ingredients, bring to a boil and simmer for 45 minutes. Cool the mixture and pour it over the fish. If the salmon is not immersed, hold it down with a crushed piece of plastic wrap. Seal. Store in the refrigerator.

STANLEY SCHULER AND ELIZABETH MERIWETHER SCHULER
PRESERVING THE FRUITS OF THE EARTH

Pickled Smelts

To make about 2 pints [1 liter]

24	smelts, cleaned, heads and tails cut off	24
2 tsp.	salt	10 ml.
about 2 cups	vinegar	about ½ liter
about 2 cups	water	about ½ liter
1	onion, stuck with 3 whole cloves	1
1	carrot, sliced	1
2	bay leaves	2
12	black peppercorns	12
4	lemon slices	4

In the buttered top section of a steamer, lay the fish flat, one above the other; sprinkle the fish with the salt, and steam over boiling water for five minutes. Then lay the fish in a deep stoneware dish, one above the other.

Boil the vinegar and water together with the remaining ingredients for about 15 minutes. Cool this mixture. Pour the mixture over the fish, cover, and refrigerate for at least 48 hours. This dish will keep perfectly fresh, under refrigeration, for up to 10 days.

JESSIE CONRAD
HOME COOKERY

Pickled Fish

Zuur van Vis

Fish prepared this way will keep for two to three weeks in the refrigerator.

To make about 6 pints [3 liters]

4 lb.	fresh turbot or cod fillets, cut into 1½-inch [4-cm.] cubes	2 kg.
2 quarts	water	2 liters
2½ cups	salt	625 ml.
¼ cup	vegetable oil	50 ml.
2 tsp.	ground turmeric	10 ml.
4	thin slices fresh ginger or 1 tsp. [5 ml.] ground ginger	4
5	small onions, thinly sliced	5
1	hot red chili, stemmed, seeded and thinly sliced	1
1 tbsp.	prepared mustard	15 ml.
2	garlic cloves, thinly sliced	2
2 quarts	vinegar	2 liters

Heat the water and salt together until the salt is thoroughly dissolved. Cool. Put the fish cubes in a glass or stoneware crock, and cover them with the cooled brine. Place a clean cloth on top of the fish, then a round wooden board that fits inside the crock, then a stone or other heavy weight. The board should keep the fish immersed in the brine. Set the crock aside overnight.

The next day, remove the fish from the brine and dry it with a cloth or kitchen towel. In a saucepan, heat the oil and sauté the fish pieces gently for a few minutes. Remove them from the pan with a slotted spoon, let them cool on a plate and place them in glass jars.

Meanwhile, combine the turmeric, ginger and onions in a mortar, and pound them to a paste. Mix with the chili, mustard and garlic. Sauté the mixture for one minute in the saucepan used for the fish, stirring all the time. Add the vinegar, bring to a boil and then let the mixture cool.

Pour the cooled mixture over the fish in the jars and put on the lids. Refrigerate. The vinegar mixture should cover the fish completely. The fish will be ready to eat in two days.

HUGH JANS
BISTRO KOKEN

Spiced and Pickled Shrimp

To make about 6 pints [3 liters]

5 lb.	fresh shrimp, peeled	2½ kg.
5 quarts	water	5 liters
1 quart	vinegar	1 liter
½ cup	salt	125 ml.
12	bay leaves	12
2 tbsp.	crushed hot chilies	30 ml.
1 tbsp.	mustard seeds	15 ml.
1 tbsp.	whole allspice	15 ml.
1 tbsp.	whole cloves	15 ml.
6	slices lemon	6
1 tbsp.	sugar	15 ml.

In an enameled or stainless-steel pan, combine 4 quarts [4 liters] of water, 2 cups [½ liter] of vinegar, the salt, 1 tablespoon [15 ml.] of chilies, 6 bay leaves, and ½ tablespoon [7 ml.] each of the mustard seeds, allspice and cloves, and simmer for 10 minutes. Bring the liquid to a boil and add the shrimp. Simmer the shrimp for five minutes, drain them—discarding the cooking liquid—and let them cool. Pack the shrimp into clean jars, adding to each jar a fresh bay leaf, a slice of lemon, ½ teaspoon [2 ml.] of chilies, and ¼ teaspoon [1 ml.] each of mustard seeds, allspice and cloves. Fill the jars with a solution made from the remaining 1 quart [1 liter] of water, 2 cups of vinegar and the sugar. Close the jars and refrigerate. Use within four to six weeks.

UNIVERSITY OF CALIFORNIA COOPERATIVE EXTENSION
MARINE BRIEFS

Standard Preparations

Simple Syrup

This recipe produces a light syrup—the kind most often used for canning fruit. For a medium syrup, mix 2¾ cups [675 ml.] of sugar with 3½ cups [875 ml.] of water. For a heavy syrup, increase the proportions to 3½ cups of sugar and 3 cups [¾ liter] of water. Heavy syrup is most often used for freezing fruit. You will need between 1 and 2 cups [¼ and ½ liter] of syrup for each quart [1 liter] of prepared fruit—the proportion varying with the density of the fruit.

To make 5 cups [1 ¼ liters]

2 cups	sugar	½ liter
4 cups	water	1 liter

Place the sugar and water in a pan. Let the sugar soak for 10 minutes, then stir over low heat to dissolve it. Increase the heat and bring the mixture briefly to a boil. Keep the syrup hot until needed.

Apple Pectin

The techniques of making and testing pectin are shown on pages 44-45. Two thirds cup [150 ml.] of this pectin syrup is sufficient to set 1 quart [1 liter] of low-pectin fruit juice.

To make 4 pints [2 liters]

10 lb.	apples or crab apples, including the peel, core and seeds, thinly sliced (about 9 quarts [9 liters])	4½ kg.
about 2½ quarts	water	about 2½ liters

Place the apples in a large enameled, tinned or stainless-steel pan, and barely cover them with water. Cover the pan, bring the mixture to a boil and simmer for about 20 minutes, or until the apples are soft. Then pour the contents of the pan into a jelly bag, and let the juice drip through for 24 hours.

Pour the strained juice back into the pan and boil it uncovered for 20 minutes, or until it is reduced by about half its original volume and is thick and syrupy.

Strain the syrup into a bowl through a colander lined with a double layer of cheesecloth or muslin. Immediately ladle the strained syrup into warm jars and cover the jars. Process the jars in a boiling water bath for 15 minutes, or cool the jars and then freeze the syrup.

Berry Juice

Any soft, ripe berry can be used: blackberries, blueberries, boysenberries, currants, elderberries, gooseberries, huckleberries, loganberries, raspberries or strawberries.

To make 2 pints [1 liter]

3 quarts	berries	3 liters
¼ to 1 cup	sugar	50 to 250 ml.

Place the berries in a deep enameled, tinned or stainless-steel pan. Crush the berries with a wooden potato masher or heavy wooden spoon to release their juice. Stirring frequently, simmer over low heat until the berries are soft—about 10 minutes. Pour the berries into a jelly bag set over a ceramic or glass bowl, and let the juice drip through without pressing on the berries. Pour the juice into the pan, stir in sugar to taste, and heat the mixture to 190°F. [88°C.]. Pour the juice into jars, cover and process it.

Cranberry juice. Combine 1 quart [1 liter] of berries with 1 quart of water and, stirring often, cook the mixture over high heat until the berries begin to burst—about three minutes. Crush any whole berries, then pour the mixture into a jelly bag. Sweeten the juice to taste, heat to 190°F. [88°C.], pour into jars, cover and process *(pages 30-31)*.

Fruit Nectar

This nectar can be made from apricots, cherries, peaches or plums. To prepare the fruit, halve and pit it. For nectar with a slightly piquant flavor, crack a few of the fruit pits to remove their kernels and add the kernels to the fruit-water mixture. Before serving the nectar, dilute it with an equal amount of ice water.

To make about 3 pints [1 ½ liters]

1 quart	prepared fruit	1 liter
1 cup	water	¼ liter
1 cup	sugar	¼ liter
1 tbsp.	fresh lemon juice	15 ml.

In an enameled, tinned or stainless-steel pan, bring the fruit and water to a boil over high heat, stirring occasionally. Reduce the heat to low, cover, and simmer until the fruit is tender—about 20 minutes. Purée the fruit through a food mill or sieve set over a bowl. Return the puréed fruit to the pan, stir in the sugar and lemon juice, and heat the mixture until the sugar dissolves. Pour the nectar into hot jars. Cover and process *(pages 30-31)*.

Fruit Sauce

Apples, apricots, peaches or pears can be used for this recipe. To prepare the fruits, quarter and core the apples or pears, or halve and pit the apricots or peaches.

To make about 4 pints [2 liters]

3 quarts	prepared fruit	3 liters
3 cups	water	¾ liter
¾ cup	sugar	175 ml.
	ground cinnamon (optional)	
	grated nutmeg (optional)	

Place the prepared fruit with the water in an enameled, tinned or stainless-steel pan. Stirring occasionally, bring to a boil over high heat. Reduce the heat to low, cover the pan and simmer for about 20 minutes, or until the fruit is tender. Purée the fruit through a food mill or sieve set over a bowl. Return the puréed fruit to the pan, stir in the sugar and add spices, if desired, to taste. Heat the mixture to boiling. Pour the sauce into hot jars, leaving ½ inch [1 cm.] headspace. Cover the jars and process.

Chunky fruit sauce. Peel and core apples or pears; skin, halve and pit apricots or peaches. Cut the fruit into eighths. Cook the fruit with the water until tender, stir in the sugar and spices—if using—and ladle the sauce into jars. Cover them and process *(pages 30-31)*.

Herb Jelly

This jelly may be flavored with any desired herb—tarragon, mint, thyme—or with whole spices such as cloves, ginger root, or cinnamon sticks.

To make about 10 cups [2 ½ liters]

6 lb.	tart apples, coarsely chopped (about 6 quarts [6 liters])	2¾ kg.
	sugar	
20	sprigs fresh sage	20

Place the apples in an enameled, tinned or stainless-steel pan, and add enough water to just cover them. Stirring occasionally, cook over medium heat for 30 minutes, or until the apples are soft and pulpy. Pour the apples into a jelly bag set over a bowl. Let the juice drip, without squeezing the bag, for one day or overnight.

Measure the juice into a pan, and add 2 cups [½ liter] of sugar for each 2½ cups [625 ml.] of juice. Stir together over low heat until the sugar dissolves. Tie the sage sprigs in a cheesecloth or muslin bag, and add to the pan. Bring the juice to a boil and simmer until it reaches the jelling point. Remove the herb bag and ladle the jelly into jars at once.

Basic Meat Stock

This stock may be made, according to your taste and recipe needs, from beef, veal, pork or chicken—or a combination of these meats. For the beef, use such cuts as shank, short ribs, chuck and oxtail; for the veal, use neck, shank and rib tips; for the pork, use hocks, Boston shoulder and back ribs; for the chicken, use backs, necks, wings and carcasses. Adding gelatinous elements such as calf's feet, pig's feet or fresh pork rind will make the finished stock set to a clear, firm jelly that can serve as an aspic if prepared carefully enough.

To make about 2 quarts [2 liters]

4 to 5 lb.	meat, bones, and trimmings of beef, veal, pork or chicken	2 to 2½ kg.
1 lb.	pig's, calf's or chicken feet, pig's ears, or fresh pork rind (optional)	½ kg.
3 to 4 quarts	water	3 to 4 liters
4	carrots	4
2	large onions, 1 stuck with 2 or 3 whole cloves	2
1	celery rib	1
1	leek, split and washed	1
1	large bouquet garni	1

Put the pieces of bone on a rack in the bottom of a heavy stockpot, and place the meat and trimmings on top of them. Add cold water to cover by 2 inches [5 cm.]. Bring to a boil over low heat, starting to skim before the liquid comes to a boil—which may take an hour. Keep skimming, occasionally adding a glass of cold water, until no scum rises—this may take 30 minutes. Do not stir, lest you cloud the stock.

Add the vegetables and bouquet garni to the pot, pushing them down into the liquid so that everything is submerged. Continue skimming until the liquid again reaches a boil. Reduce the heat to very low, partially cover the pan, and cook undisturbed at a bare simmer for two hours if you are using only chicken trimmings, otherwise for five hours.

Skim off the surface fat. Then strain the stock by pouring the contents of the pot through a colander into a large bowl or clean pot. Discard the bones and meat trimmings, vegetables and bouquet garni. Cool the strained stock and remove the last traces of fat from the surface with a folded paper towel. If there is any residue at the bottom of the container after the stock cools, pour the clear liquid slowly into another container and discard the sediment.

Refrigerate the stock if you do not plan to use it immediately; it will keep safely for three to four days. To preserve the stock longer, refrigerate it for only 12 hours—or until the last bits of fat solidify on the top—then scrape off the fat and warm the stock enough so that it may be poured into four or five pint-sized [½-liter] freezer containers. Be sure to leave room in the containers to allow for expansion, and tightly cover the containers. The frozen stock will keep for six months while you draw on the supply as necessary.

Brine

To make 4 quarts [4 liters]

4 quarts	water	4 liters
3 cups	coarse salt	¾ liter
2¾ cups	dark brown sugar	675 ml.
1	bay leaf	1
1	sprig thyme	1
10	juniper berries, crushed	10
10	peppercorns, crushed	10

Combine all of the ingredients and bring to a boil. Boil for five minutes. Let the brine cool before using.

Corned Meat

Any tough cut of meat can be used for this recipe: beef round, chuck, flank, plate or brisket; pork Boston shoulder, picnic shoulder, or blade end; lamb shoulder or neck. Goose or duck, cut into serving pieces rather than slices, also can be corned by this method.

To make 5 pounds [2½ kg.]

5 lb.	meat, sliced 1 inch [2½ cm.] thick, pricked all over with a metal skewer	2½ kg.
1 cup	coarse salt	¼ liter
4 quarts	brine *(recipe, above)*	4 liters

Rub the slices of meat with salt. Put a layer of salt in a large glass or ceramic bowl, and add a layer of meat. Alternate layers of salt and meat, ending with a layer of salt. Cover, and place the bowl in the refrigerator for 24 hours.

Make the brine, strain, and let it cool. Rinse the meat under cold running water to remove the salt and the coagulated juices clinging to the slices. Place the meat in a clean bowl and cover it with the brine. Set an inverted plate on top of the meat, and weight the plate with a jar full of water to keep the meat submerged. Refrigerate for 24 hours, then use a spoon to lift off any scum that has risen to the surface. Refrigerate again and repeat the skimming daily. The meat will be ready in nine or 10 days.

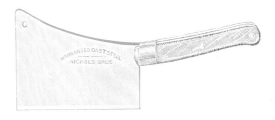

Recipe Index

All recipes in the index that follows are listed by the English title except in cases where a dish of foreign origin, such as rillettes, is universally recognized by its source name. Entries are organized in separate categories by major ingredients specified in the recipe titles. Foreign recipes are listed under the country or region of origin. Recipe credits appear on pages 174-176.

General Index/ Glossary

Included in this index to the cooking demonstrations are definitions, in italics, of special culinary terms not explained elsewhere in this volume. The Recipe Index begins on page 168.

Recipe Credits

The sources for the recipes in this volume are shown below. Page references in parentheses indicate where the recipes appear in the anthology.

Adam, Hans Karl, *Das Kochbuch aus Schwaben.* © Copyright 1976 by Verlagsteam Wolfgang Hölker. Published by Verlag Wolfgang Hölker, Münster. Translated by permission of Verlag Wolfgang Hölker(126).
American Honey Institute, *Old Favorite Honey Recipes.* Copyright 1945(103, 148).
Amicale des Cuisiniers et Pâtissiers Auvergnats de Paris, *Cuisine D'Auvergne.* © 1979 Denoël-Paris. Published by Éditions Denoël, Paris. Translated by permission of Éditions Denoël(111).
Armisen, Raymond and André Martin, *Les Recettes de la Table Niçoise.* © Librairie Istra 1972. Published by Librairie Istra, Strasbourg. Translated by permission of Librairie Istra(101, 147).
The Art of Cookery, Made Plain and Easy. By a Lady. The Sixth Edition, 1758(123).
Aylett, Mary, *Country Fare.* © Mary Aylett 1956. Published by Odhams Press Limited, London 1956. By permission of David Higham Associates Ltd., London, for the author(96).
The Ball Blue Book. Copyright 1979. Published by Ball Corporation, Muncie, Indiana. By permission of Ball Corporation(106).
Beard, James, *Delights and Prejudices.* Copyright © 1964 by James Beard. By permission of Atheneum Publishers(159).
Beard, James, Milton Glaser, and Burton Wolf, *The Garden-to-Table Cookbook.* Copyright 1976 by Beard Glaser Wolf Ltd. Published by McGraw-Hill Book Company. By permission of McGraw-Hill Book Company(124).
Beeton, Mrs. Isabella, *The Book of Household Management.* First published in 1861. Reproduced in facsimile by Jonathan Cape Ltd., London(159).
Bernardin Home Canning Guide. Published by Bernardin, Inc., Evansville, Inc. © 1975, All Rights Reserved. By permission of Bernardin, Inc.(92).
Blencowe, Ann, *The Receipt Book of Ann Blencowe (A.D. 1694).* Published by Guy Chapman, The Adelphi, London 1925(135, 162).
Boulestin, X. Marcel, *A Second Helping or More Dishes for English Homes.* Published by William Heinemann Ltd., London 1928. By permission of A. D. Peters and Co., Ltd., London(114, 118).
Breteuil, Jules, *Le Cuisinier Européen.* Published by Garnier Frères Libraires-Éditeurs c. 1860(95).
British Columbia Women's Institutes, *Adventures in Cooking.* Published by British Columbia Women's Institutes, British Columbia 1958. By permission of British Columbia Women's Institutes(140, 142).
Brobeck, Florence, *Old-Time Pickling and Spicing Recipes.* Copyright © 1953 by Florence Brobeck. Published by Gramercy Publishing Company, New York. By permission of William Morrow & Company, New York(127, 138).
Brown, Marion, *Pickles and Preserves* (Funk & Wagnalls). Copyright © 1955 by Wilfred Funk, Inc. By permission of Harper & Row, Publishers, Inc.(150).
Brownstone, Cecily, *Cecily Brownstone's Associated Press Cookbook.* Reprinted with permission, copyright 1972. Published by David McKay Co., Inc.(146).
The Buckeye Cookbook: Traditional American Recipes. As published by the Buckeye Publishing Co. 1883. Republished by Dover Publications, Inc., New York 1975(94, 122, 157).
Butel, Jane, *Jane Butel's Tex-Mex Cookbook.* Copyright © 1980 by Jane Butel. By permission of Harmony Books(112).
Byron, May, *May Byron's Jam Book.* © Hodder and Stoughton 1916. Published by Hodder and Stoughton, London. By permission of Hodder and Stoughton Limited(98, 104, 150).

Callahan, Genevieve, *The New California Cook Book.* Copyright 1946 by Genevieve Callahan. Copyright © 1955 by Genevieve Callahan. By permission of William Morrow & Company(110).
Cameron-Smith, Marye, *The Complete Book of Preserving.* © Marshall Cavendish Limited 1976. Published by Marshall Cavendish Publications Limited, London. By permission of Marshall Cavendish Publications Limited(97, 114).
Castignac, Huguette, *La Cuisine Occitane.* © Solar 1973. Published by Solar, Paris. Translated by permission of Solar(103, 107, 113).
Cecchini, Tina, *Les Conserves de Fruits et Légumes.* © Éditions De Vecchi S.A. — Paris, 1976. Published by Éditions De Vecchi. Translated by permission of Éditions De Vecchi S.A.(108, 110, 144, 155).
Chortanova, Sonya, *Nasha Kichniya.* Published by Naukii Izkustvo, Sofia 1955. Translated by permission of Jusautor Copyright Agency, Sofia(110).
Cobbett, Anne, *The English Housekeeper.* Originally published by A. Cobbett, Strand, 5th edition, 1851. Reprinted in facsimile by EP Publishing Limited. Copyright © 1973 by EP Publishing Limited, Wakefield, Yorkshire(120, 123, 124).
Comelade, Éliane Thibaut, *La Cuisine Catalane.* © Éditions CLT J. Lanore. Published by Éditions Jacques Lanore, Paris 1978. Translated by permission of Éditions Jacques Lanore(100, 147).
Conrad, Jessie, *Home Cookery.* Published by Jarrolds Publishers Ltd., London 1936. By permission of the Hutchinson Publishing Group Ltd., London(164).
Conran, Caroline, *British Cooking.* © Caroline Conran Ink Ltd. 1978. First published by Park Lane Press, London 1978. By permission of The Rainbird Publishing Group Ltd., London, and André Deutsch Limited, London(100, 139, 140).
Corbitt, Helen, *Helen Corbitt's Cookbook.* Copyright © 1957 by Helen L. Corbitt. Published by Houghton Mifflin Company, Boston. By permission of Houghton Mifflin Company(111).
Le Cordon Bleu. Published by Le Cordon Bleu de Paris, 1932. Translated by permission of Le Cordon Bleu de Paris(105, 108).
Craig, Elizabeth, *Court Favourites.* First published by André Deutsch Limited, London 1953. By permission of André Deutsch Limited(123). *The Scottish Cookery Book.* © Elizabeth Craig, 1956. Published by André Deutsch Limited, London 1956. By permission of André Deutsch Limited(104, 115).
Il Cuoco Piemontese Ridotto all'Ultimo Gusto. Published in Milano 1828(102, 152).
Cutler, Carol, *Haute Cuisine for Your Heart's Delight.* Copyright © 1973 by Carol Cutler. By permission of Clarkson N. Potter, Inc.(133). *The Six Minute Soufflé and Other Culinary Delights.* Copyright © 1976 by Carol Cutler. By permission of Clarkson N. Potter, Inc.(116).
Czerny, Z. and M. Strasburger, *Zwienie Rodziny.* Copyright by Zofia Czerny. Originally published by PWE. Translated by permission of Agencja Autorska, Warsaw(104).
The Daily Telegraph, *400 Prize Recipes for Practical Cookery.* © by The Daily Telegraph. Published by The Daily Telegraph, London, c. 1950. By permission of The Daily Telegraph(118, 122, 144). *New Dishes from The Daily Telegraph.* Published by H. A. & W. L. Pitkin, Ltd., London, for The Daily Telegraph, c. 1953. By permission of The Daily Telegraph, London(101).
David, Elizabeth, *Spices, Salt and Aromatics in the English Kitchen.* Copyright © Elizabeth David, 1970. Published by Penguin Books Ltd., London. By permission of Penguin Books Ltd.(112).
Davidson, Alan, *North Atlantic Seafood.* Copyright © 1979 by Alan Davidson. By permission of Viking Penguin Inc.(163).
de Groot, Roy Andries, *The Auberge of the Flowering Hearth.* Copyright © 1973 by Roy Andries de Groot. Published by The Bobbs-Merrill Company, Inc., Indianapolis/New York. By permission of Robert Cornfield, as Agent for the author(130).

de Lazarque, E. Auricoste, *Cuisine Messine.* Published by Sidot Frères, Libraires-Éditeurs, Nancy, 1927. Reprinted by Laffitte Reprints, Marseilles, 1979. Translated by permission of Laffitte Reprints(126, 149).
DeVore, Sally and Thelma White, *The Appetites of Man.* Copyright © 1977 by Sally DeVore and Thelma White. Published by Anchor Press, Doubleday and Co. Inc. Originally published under the title of "Dinner's Ready!" in 1977. By permission of the Merritt Literary Agency for the author(102).
Elkon, Juliette and Elaine Ross, *Menus for Entertaining.* Copyright © 1980. Permission by Hastings House, Publishers(106).
Escudier, Jean-Noël, *La Veritable Cuisine Provençale et Niçoise.* Published by U.N.I.D.E., Paris 1974. Translated by permission of U.N.I.D.E.(162).
Famularo, Joe and Louise Imperiale, *The Festive Famularo Kitchen.* Copyright © 1977 by Joe Famularo and Louise Imperiale. Published by Atheneum Publishers, New York 1977. By permission of Atheneum Publishers(131, 136).
Filippini, Maria Nunzia, *La Cuisine Corse.* Published by Société d'Éditions Serena, Ajaccio 1978. Translated by permission of Société d'Éditions Serena(153).
Firth, Grace, *A Natural Year.* Copyright © 1973 by Grace Firth. Published by Simon and Schuster, New York. By permission of Simon and Schuster, a division of Gulf and Western Corporation(115).
Flexner, Marion, *Out of Kentucky Kitchens.* Copyright 1949 by Marion Flexner. Published by Franklin Watts, Inc., New York. By permission of Franklin Watts, Inc.(107, 117, 121).
Foods of the World, *American Cooking: The Eastern Heartland.* © 1971 Time Inc. Published by Time-Life Books, Alexandria, Virginia(125).
Frere, Catherine Frances (Editor), *The Cookery Book of Lady Clark of Tillypronie.* Published by Constable & Company Ltd., London 1909(95, 109, 148).
Garrett, Blanche Pownall, *Canadian Country Preserves & Wines.* Copyright © 1974 by Blanche Pownall Garrett. Published by James Lewis & Samuel, Publishers, Toronto 1974. By permission of James Lorimen and Company Limited, Toronto(137).
Gavotti, Erina (Editor), *Millericette.* © Copyright 1965 by Garzanti editore. Published by Aldo Garzanti editore, Milan 1966. Translated by permission of Aldo Garzanti editore sp.a.(146, 148).
Gerber, Hilda, *Traditional Cookery of the Cape Malays.* Published by A. A. Balkema/Amsterdam/Cape Town 1959. By permission of A. A. Balkema, Rotterdam(153).
Gewanter, Vera and Dorothy Parker, *Home Preserving Made Easy.* Copyright © 1975 by Vera Gewanter and Dorothy Parker. By permission of Viking Penguin, Inc.(97, 163).
Gibbons, Euell, *Stalking the Wild Asparagus.* Reprinted with permision. Copyright 1962. Published by David McKay Co., Inc.(105).
Gins, Patricia, *Great Southwest Cooking Classic.* Copyright The Albuquerque Tribune, 1977. By permission of The Albuquerque Tribune(141).
Godard, Misette, *Le Temps des Confitures.* Copyright © Éditions Seghers, Paris 1977. Published by Éditions Seghers, Paris. Translated by permission of Éditions Seghers/Éditions Robert Laffont, Paris(103, 107).
Grigson, Jane, *The Art of Making Sausages, Pâtés and Other Charcuterie.* Copyright © 1967 by Jane Grigson. Reprinted by permission of Alfred A. Knopf, Inc.(136). *Good Things.* Copyright © 1968, 1969, 1970, 1971 by Jane Grigson. Copyright © 1971 by Alfred A. Knopf, Inc. By permission of Alfred A. Knopf, Inc.(160).
Groff, Betty and José Wilson, *Good Earth & Country Cooking.* Published by Stackpole Books, Harrisburg, Pennsylvania, 1974. By permission of Stackpole Books(129).
Guthrie, Ruby Charity Stark and Jack Stark Guthrie, *A Primer for Pickles.* Copyright 1974 by Ruby Charity Stark Guthrie and Jack Stark Guthrie. Published by 101 Productions, San Francisco. By permission of 101 Productions(133, 142).

Hadzhiyski, T., D. Donkov, N. Pekavchev, P. Koen, M. Tzolova, *Domashno Konservirane.* © by the authors. Published by Zimizdat, Sofia 1972. Translated by permission of Jusautor Copyright Agency, Sofia(93, 94).

Haitsma Mulier-Van Beusekom, C. A. H. (Editor), *Culinaire Encyclopédie.* Published by Elsevier © 1957. Revised edition 1974 by N. V. Uitgeversmaatschappij Elsevier Nederland and E. H. A. Nakken-Rovekamp. Translated by permission of B. V. Uitgeversmaatschappij Elsevier Focus, Amsterdam(160).

The Hammond-Harwood House Association, *Maryland's Way.* Copyright 1966. By permission of The Hammond-Harwood House Association, Annapolis, Maryland(154).

Hellermann, Dorothee V., *Das Kochbuch aus Hamburg.* © copyright 1975 by Verlagsteam Wolfgang Hölker. Published by Verlag Wolfgang Hölker, Münster. Translated by permission of Verlag Wolfgang Hölker(134, 155).

Heritage, Lizzie, *Cassell's Universal Cookery Book.* Published by Cassell and Company, Limited, London 1901(120, 122).

Hertzberg, Ruth, Beatrice Vaughn and Janet Greene, *Putting Food By.* © 1973, 1974, 1975 by The Stephen Greene Press, Brattleboro, Vermont. By permission of The Stephen Greene Press(148).

Hewitt, Jean, *The New York Times Natural Foods Cookbook.* Copyright © 1971 by Jean Hewitt. By permission of Times Books, a Division of Quadrangle/The New York Times Book Co.(108). *The New York Times New England Heritage Cookbook.* Copyright © 1977 by the New York Times Company. By permission of G. P. Putnam's Sons(105, 150).

Hume, Rosemary and Muriel Downes, *Jams, Preserves and Pickles.* © 1972 with the permission of Contemporary Books, Inc., Chicago(156).

Hutchinson, Peggy, *Grandma's Preserving Secrets.* © W. Foulsham & Co., Ltd. 1978. Published by W. Foulsham & Co., Ltd., London. By permission of W. Foulsham & Co. Ltd.(102, 136). *Peggy Hutchinson's Preserving Secrets.* © W. Foulsham & Co., Ltd. Published by W. Foulsham & Co. Ltd., London, c. 1930. By permission of W. Foulsham & Co. Ltd.(96).

Hutchison, Ruth, *The Pennsylvania Dutch Cook Book.* Copyright 1948 by Harper & Row, Publishers, Inc. By permission of the publisher(140).

Innes, Jocasta, *The Country Kitchen.* Copyright © 1979 by Jocasta Innes. Published by Frances Lincoln Publishers Limited, London in association with Weidenfeld and Nicolson Limited, London. By permission of Frances Lincoln Publishers Limited(118).

Jans, Hugh, *Bistro Koken.* © Unieboek BV/C.A.J. van Dishoeck, Bussum, Holland. Translated by permission of Unieboek BV/C.A.J. van Dishoeck(164).

Käkönen, Ulla, *Natural Cooking the Finnish Way.* Copyright © 1974 by Ulla Käkönen. Published by Quadrangle/The New York Times Book Company, New York. By permission of Times Books, a division of Quadrangle/The New York Times Book Company(128, 133).

Karsenty, Irène and Lucienne, *La Cuisine Pied-Noir.* (Cuisines du Terroir). © 1974, by Éditions Denoël, Paris. Published by Éditions Denoël. Translated by permission of Éditions Denoël(112).

Kolder, E., *Kuchnia Slaska.* Published by Profil, Ostrava 1978. Translated by permission of the heir to the author, c/o DILIA Theatrical and Literary Agency, Prague(117).

Krochmal, Connie and Arnold, *Caribbean Cooking.* Copyright © 1974 by Connie and Arnold Krochmal. Published by Quadrangle/The New York Times Book Co., New York. By permission of Times Books, A Division of Quadrangle/The New York Times Book Co.(99, 129, 137).

Krüger, Arne and Annette Wolter, *Kochen Heute.* © by Gräfe und Unzer GmbH, München. Published by Gräfe und Unzer GmbH, Munich, 1972. Translated by permission of Gräfe und Unzer GmbH(150).

Kurdzhieva, Lidiya (Editor), *Domashno Prigotvyane na Zimnina.* © by the author. Published by Znanie, Sofia, c. 1925. Translated by permission of Jusautor Copyright Agency, Sofia(112, 119).

Lal, Premila, *Premila Lal's Indian Recipes.* © Premila Lal. Published by Rupa & Co., Calcutta, 1974. By permission of Rupa & Co.(137, 138).

Levinson, Leonard Louis, *The Complete Book of Pickles & Relishes.* Copyright © 1965 by Leonard Louis Levinson. Published by Hawthorn Books, Inc., Publishers/New York. By permission of Hawthorn Books, a division of Elsevier-Dutton Publishing Co., Inc.(133, 135, 159).

Levy, Paul, *Habitat Cook's Diary 1980.* Copyright © Conran, A. L., 1980. Published by Habitat Designs Limited, London. By permission of Habitat Designs Limited(129).

Leyel, Mrs. C. F., *The Complete Jam Cupboard.* Published by Geroge Routledge & Sons Ltd., London. By permission of Routledge & Kegan Paul Ltd., London(109).

Linnich, Eike, *Das Grosse Einmachbuch.* © 1977 by Mosaik Verlag GmbH, München. Published by Mosaik Verlag GmbH, Munich. Translated by permission of Mosaik Verlag GmbH(99, 115, 152, 157).

Louisiana State University, *Louisiana Cooperative Extension Publication No. 1568.* By permission of Louisiana Cooperative Extension Service(99).

Mabey, David & Rose, *Jams, Pickles & Chutneys.* Copyright © 1975 David and Rose Mabey. Published by Penguin Books Limited, London/Macmillan Limited, London. By permission of Penguin Books Ltd.(114, 116, 128).

Magyar, Elek, *Kochbuch für Feinschmecker.* © Dr. Magyar Bálint. © Dr. Magyar Pál. Originally published in 1967 under the title "Az Inyesmester Szakacs Konyve" by Corvina Verlag, Budapest. Translated by permission of Artisjus, Literary Agency, Budapest(134).

Mann, Gertrude, *The Apple Book.* Published by André Deutsch Limited, London 1953. By permission of André Deutsch Limited(100, 125). *Berry Cooking.* Published by André Deutsch Limited, London 1954. By permission of André Deutsch Limited(104, 120).

Markovic, Spasenija-Pata (Editor), *Veliki Narodni Kuvar.* Copyright by the author. Published by Narodna Knjiga, Belgrade 1979. Translated by permission of Jogoslovenska Autorska Agencija, Belgrade, for the heir to the author(102, 116).

Mathiot, Ginette, *Je Sais Faire Les Conserves.* Copyright 1948, by Éditions Albin Michel, Paris. Published by Éditions Albin Michel, Paris. Translated by permission of Éditions Albin Michel(93, 95, 109, 112).

Meighn, Moira, *The Magic Ring for the Needy & Greedy.* Published by Oxford University Press, London 1936. By permission of Oxford University Press(99).

Menon, *La Cuisinière Bourgeoise.* Paris 1746(152).

Miller, Carey D., Katherine Bazore and Mary Bartow, *Fruits of Hawaii.* Copyright © 1955, 1965 by the University of Hawaii Press. Permission by The University Press of Hawaii(105).

Mitchell, Alice Miller, *Oriental Cookbook.* Copyright 1954 by Alice Miller Mitchell. Published by Rand McNally & Company(155).

Montagné, Prosper, *The New Larousse Gastronomique.* English text © 1977 by Hamlyn Publishing Group Limited. Published by Crown Publishers, Inc. By permission of Crown Publishers, Inc.(130).

Nichols, Nell B. and Kathryn Larson (Editors), *Farm Journal's Freezing & Canning Cookbook.* Copyright © 1963, 1964, 1973, 1978 by Farm Journal Inc. Published by Doubleday & Company, Inc., New York 1978. By permission of Farm Journal, Inc., Philadelphia(109, 127, 132, 156).

Nignon, Edouard (Editor), *Le Livre de Cuisine de L'Ouest-Eclair.* Published by L'Ouest-Éclair, Rennes 1941. Translated by permission of Société d'Éditions Ouest-France, Rennes(111, 160, 162).

Norberg, Inga, *Good Food from Sweden.* Published by Chatto & Windus, London, 1935. By permission of Curtis Brown Ltd., London, Agents for the author(135).

Oberosler, Giuseppe (Editor), *Il Tesoretto della Cucina Italiana.* © 1980 Hoepli-Milano. Published by Editore Ulrico Hoepli, Milan 1948, 1980. Translated by permission of Ulrico Hoepli s.p.a.(127, 131, 139).

Odell, Olive, *Preserves and Preserving.* © W1 Books Ltd., 1978. Published by Macdonald Educational Ltd., London. By permission of Macdonald Educational Ltd.(145).

Olney, Judith, *Summer Food.* Copyright © 1978 by Judith Olney. Published by Atheneum Publishers, Inc., New York. By permission of Atheneum Publishers, Inc.(130).

Ortiz, Elisabeth Lambert, *The Complete Book of Mexican Cooking.* Copyright © 1967 by Elisabeth Lambert Ortiz. By permission of the publisher, M. Evans and Company, Inc., New York, New York(118).

Paradissis, Chrissa, *The Best Book of Greek Cookery.* Copyright © 1976 P. Efstathiadis & Sons. Published by Efstathiadis Group, Athens, 1976. By permission of P. Efstathiadis & Sons S.A.(119).

Parke, Gertrude, *Going Wild in the Kitchen.* Copyright by Gertrude Parke 1965. Published by David McKay Company, Inc. By permission of Gertrude Parke(116).

Petits Propos Culinaires 3. November, 1979. Copyright © Prospect Books 1979. Published by Prospect Books, London and Washington D.C. By permission of the publisher(98, 106, 154).

Petits Propos Culinaires 5. May, 1980. Copyright © Prospect Books 1980. Published by Prospect Books, London and Washington D.C. By permission of the publisher(110).

Philippon, Henri, *Cuisine du Quercy et du Périgord.* © 1979, by Éditions Denoël, Paris. Published by Éditions Denoël, Paris. Translated by permission of Éditions Denoël(98).

The Pleasures of Cooking Vol. 1 No. 12. Copyright © 1979 by Cuisinart® Cooking Club, Inc. Published by Cuisinart® Cooking Club, Inc., Connecticut. By permission of Cuisinart® Cooking Club, Inc.(151).

Il Re dei Cuochi. Published by Adriano Salani, Editore, Florence 1891(111, 149).

Richardson, Mrs. Don (Editor), *Carolina Low Country Cookbook of Georgetown, South Carolina.* © 1947 Mrs. Don Richardson, Georgetown, S.C. Printed by Walker, Evans and Cogswell Co., Charleston, South Carolina 1963, 1975 for Women's Auxiliary, Prince George, Winyah, Protestant Episcopal Church, Georgetown, S.C. By permission of Mrs. Don Richardson and Ruth O'Bell(138).

Ripoll, Luis, *Cocina de las Baleares.* © by Luis Ripoll. Published in Majorca 1974. Translated by permission of Luis Ripoll(103).

Rogledi, Gianna Montecucco, *Sotto Vetro.* Published by Longanesi Milan 1973. Translated by permission of Longanesi & Co., Milan(141, 143, 144).

Rombauer, Irma S. and Marion Rombauer Becker, *Joy of Cooking.* Copyright © 1931, 1936, 1941, 1942, 1943, 1946, 1951, 1952, 1953, 1962, 1963, 1964, 1975 by The Bobbs-Merrill Company, Inc. Published by The Bobbs-Merrill Company, Inc., Indianapolis/New York 1975. By permission of The Bobbs-Merrill Company, Inc.(94).

Rossi, Emmanuele (Editor), *La Vera Cuciniera Genovese.* Published by Casa Editrice Bietti S.p.A., Milan, 1973. Translated by permission of Casa Editrice Bietti S.p.A.(125, 131).

Sarrau, José, *Mi Recetario de Cocina.* Published by Imprenta Prensa Española, Madrid. Translated by permission of the author(113).

Scheibler, Sophie Wilhelmine, *Allgemeines deutsches Kochbuch für alle Stände.* Published in Leipzig, 1896(121, 148).

Schuler, Elizabeth, *Mein Kochbuch.* © Copyright 1948 by Schuler-Verlag, Stuttgart-N, Lenzhalde 28. Published by Schuler Verlagsgesellschaft. Translated by permission of Schuler Verlagsgesellschaft(95, 141).

Schuler, Stanley and Elizabeth Meriweather Schuler, *Preserving the Fruits of the Earth.* Copyright © 1973 by Stanley Schuler and Elizabeth Meriweather Schuler. By permission of the publisher, The Dial Press(112, 164).

Scott, David, *The Japanese Cookbook.* © David Scott 1978. Published by Barrie and Jenkins Ltd., London. By permission of the Hutchinson Publishing Group Limited, London(111).

Seranne, Ann, *The Complete Book of Home Preserving.* Copyright © 1955 by Ann Seranne. By permission of Doubleday & Company, Inc.(96, 98, 101).

Serra, Victoria, *Tia Victoria's Spanish Kitchen.* English text copyright © Elizabeth Gili, 1963. Published by Kaye & Ward Ltd., London, 1963 and Weathervane Books, New

York. Translated by permission of Elizabeth Gili from the original Spanish entitled "Sabores, Cocina del Hogar" by Victoria Serra Suñol. By permission of Kaye & Ward Ltd.(113, 114, 119).

Showalter, Mary Emma, *Mennonite Community Cookbook.* Copyright 1950, 1957, renewed 1978 by Mary Emma Showalter. By permission of Herald Press, Scottdale, Pennsylvania(157).

Snow, Jane Moss, *A Family Harvest.* Copyright © 1976 by Jane Moss Snow. Published by The Bobbs-Merrill Company, Inc., Indianapolis/New York. By permission of Collier Associates, New York, for the author(126).

Sorzio, Angelo, *The Art of Home Canning.* Copyright 1976 Leon Amiel Publisher, Inc., 31 W.46th Street, New York, N.Y. By permission of Leon Amiel Publisher, Inc.(92, 94, 107, 155).

Stoner, Carol Hupping, *Stocking Up.* © 1977 by Rodale Press, Inc. By permission of Rodale Press, Inc., Emmaus, Pennsylvania(152, 158).

Szathmáry, Louis, *American Gastronomy.* Copyright © 1974 by Louis Szathmáry. Published by Henri Regnery Company, Chicago, by special arrangement with Arno Press, Inc., New York. By permission of Arno Press Inc.(121, 122).

Tenison, Marika Hanbury, *Recipes from a Country Kitchen.* Copyright © Marika Hanbury Tenison 1978. Published by Hart-Davis, MacGibbon Ltd./Granada Publishing Ltd., Hertfordshire. By permission of A. D. Peters & Co., Ltd., London(96, 161).

Tiano, Myrette, *Les Conserves.* © Solar, 1979. Published by Solar, Paris. Translated by permission of Solar(93). *Conserves Maison.* © Solar, 1977. Published by Solar, Paris. Translated by permission of Solar (162).

Tschirky, Oscar, *The Cook Book by "Oscar" of the Waldorf.* Copyright 1896 by Oscar Tschirky. Published by The Werner Company, Chicago and New York(93, 143).

Tyree, Marion Cabell, *Housekeeping in Old Virginia.* Copyrighted by John P. Morton and Company, 1879. Published by John P. Morton and Company, 1890(120).

Tyson, Miss, *The Queen of the Kitchen.* Published by T. B. Peterson & Brothers, Philadelphia 1874(100, 106, 124).

University of California Cooperative Extension, *Marine Briefs.* By permission of Division of Agricultural Sciences, University of California(165). *Safe Methods for Preparing Pickles, Relishes & Chutneys.* By permission of Division of Agricultural Sciences, University of California(134, 156).

U.S. Department of Agriculture, *Complete Guide to Home Canning.* Copyright © 1973 by Dover Publications, Inc. Published by Dover Publications, Inc.(128).

Varenne, La, *Le Vray Cuisinier François.* Published by Pierre Mortier, Libraire, Paris 1651(97).

Vence, Céline, *Encyclopédie Hachette de la Cuisine Régionale.* © Hachette 1979. Published by Librairie Hachette, Paris. Translated by permission of Librairie Hachette(153, 154). *Le Grand Livre des Conserves, des Confitures et de la Congélation.* © Le Livre de Paris — Hachette, 1979. Published by Librairie Hachette, Paris. Translated

by permission of Librairie Hachette(142, 161).

Viard and Fouret, *Le Cuisinier Royal.* Paris, 1828(149).

Wakefield, Ruth Graves, *Toll House Tried and True Recipes.* Published by Dover Publications, Inc.(158).

Wejman, Jacqueline, *Jams & Jellies.* Copyright 1975 by Jacqueline Wejman and Charles St. Peter. Published by 101 Productions, San Francisco. By permission of 101 Productions(147).

White, Florence, *Flowers as Food.* Published by Jonathan Cape Ltd., London 1934. By permission of Jonathan Cape Ltd.(132).

White, Patricia Holden, *Food as Presents.* Copyright © 1975 by Patricia Holden White. Published by Faber and Faber Limited, London. By permission of Faber and Faber Limited(97, 126).

Witty, Helen and Elizabeth Schneider Colchie, *Better than Store-Bought.* Copyright © 1979 by Helen Witty and Elizabeth Schneider Colchie. Published by Harper & Row, Publishers, Inc., New York. By permission of Harper & Row, Publishers, Inc.(92, 108, 132, 163).

Woman's Auxiliary of Olivet Episcopal Church, *Virginia Cookery —Past and Present.* Copyright 1957 by the Woman's Auxiliary of Olivet Episcopal Church, Franconia, Virginia. By permission of the publisher(146, 153).

Woman's Institute Library of Cookery. *Jelly Making, Preserving, and Pickling.* Copyright 1918 by International Educational Publishing Company. Published by Woman's Institute of Domestic Arts and Sciences, Inc., Scranton, Pennsylvania(108, 145).

Acknowledgments

The indexes for this book were prepared by Louise W. Hedberg. The editors are particularly indebted to Lee Ann Blum, Potomac, Maryland; Dr. F. William Cooler, Dr. George Flick, Virginia Polytechnic Institute, Blacksburg; Gail Duff, Kent, England; Jocasta Innes, London; Norman Kolpas, New York; Dr. Roger McFeeters, North Carolina State University, Raleigh; Mary Norwak, Norfolk, England; Olive Odell, Worcestershire, England; Ann O'Sullivan, Majorca, Spain; Dr. R. H. Smith, Aberdeen, Scotland.

The editors also wish to thank: Mary Attenborough, Essex, England; Elizabeth Bailey, London; Cannon Seafood Inc., Washington, D.C.; Liz Clasen, London; Emma Codrington, Surrey, England; Carol Davis, U.S. Department of Agriculture, Consumer Nutrition Center, Beltsville, Maryland; Charlotte M. Dunn, University of Wisconsin, Madison; The Electrical Association for Women, London; The Electricity Council, London; Dafne Engstrom, San Francisco; Mimi Errington, London; Maggie Heinz, London; Hudson Brothers Greengrocers, Washington, D.C.; Katherine Humphrey, Mt. Morris, New York; Frederica L. Huxley, London; Brenda Jayes, London; Maria Johnson, Hertfordshire, England; Wanda Kemp-Welch, Nottingham, England; Rosemary Klein, London; Long

Ashton Research Station, Bristol, England; Pippa Millard, London; Sonya Mills, Canterbury, England; Wendy Morris, London; Dilys Naylor, Surrey, England; Carol D. Owen, Ball Corporation, Muncie, Indiana; Dr. Robert Price, University of California, Davis; Michael Schwab, London; Cathy Sharpe, Annandale, Virginia; Fiona Tillett, London; Pat Tookey, London; J. M. Turnell & Co., London; Irene M. Turner, Louisiana State University, Baton Rouge; Tina Walker, London; Claire Walsh, London; Dr. Robert C. Wiley, University of Maryland, College Park; Dr. Charles Wood, Virginia Polytechnic Institute, Blacksburg; Harry Yoshimura, Mutual Fish Co., Inc., Seattle, Washington; Gretchen Ziesmer, Mirro Corporation, Manitowoc, Wisconsin

Picture Credits

The sources for the pictures in this book are listed below. Credits for each of the photographers and illustrators are listed by page number in sequence with successive pages indicated by hyphens; where necessary, the locations of pictures within pages are also indicated —separated from page numbers by dashes.

Photographs by Aldo Tutino: 8-11, 16-17, 25 —top, 26-37, 38 —bottom, 39 —top right and bottom, 40-42, 45 — bottom, 46-49, 50-51 —bottom, 52-55, 56 —bottom, 57, 58-59 —bottom, 65 —bottom, 66 —bottom right, 67 — top right, 69 —top right, 71, 73 —bottom, 74, 80-81, 84-85 —top, 86-90.
Other photographs (alphabetically): Tom Belshaw, 12, 18-

19 —bottom, 20, 21 —left and center, 22-23 —bottom, 24 —bottom, 56 —top, 58 —top, 59 —top left, top center, 64 —top, 68-69 —bottom, 70 —top, bottom left and center, 76 —bottom, 77 —top and bottom left, 78 — bottom, 79 —bottom left, bottom center, 85 —bottom right. Alan Duns, cover, 25 —bottom, 70 —bottom right, 76 — top, 77 —bottom right, 78 —top, 79 —top left, 82 —top, 83 —top left and bottom center. John Elliott, 4, 18-19 — top, 21 —top, 22-23 —top and center, 24 —top, 38 —top, 39 —top left, top center, 45 —top and center, 50-51 —top, 59 —top right, 60-63, 64 —bottom, 65 —top, 66 —top and bottom left, 67 —top left, center and bottom, 68 —top, 69 —top left, top center, 72, 73 —top, 79 —top center, top right, bottom right, 83 —except top left and bottom center, 84 —bottom, 85 —bottom left and center. Louis Klein, 2. Bob Komar, 82 —bottom.

Illustrations: From the Mary Evans Picture Library and private sources and *Food & Drink: A Pictorial Archive from Nineteenth Century Sources* by Jim Harter, published by Dover Publications, Inc., 1979, 93-164.

Library of Congress Cataloguing in Publication Data
Time-Life Books.
Preserving.
(The Good cook techniques & recipes)
Includes index.
1. Food —Preservation. I. Title. II. Series:
Good cook techniques & recipes.
TX601.T53 1981   641.4   80-26851
ISBN 0-8094-2906-3
ISBN 0-8094-2904-7 (retail ed.)
ISBN 0-8094-2905-5 (lib. bdg.)

Printed in U.S.A.